ENVIRONMENTAL PSYCHOLOGY

About the Author

Dinesh Nagar is working as Director M.P. Bhoj (open) University, Bhopal. Dr. Nagar, Reader and Coordinator of HRD Programme in the Department of Psychology, is on lien from Barkatullah University, Bhopal. He has a D.Phil degree in Psychology from Allahabad University. He has received his Post-Doctoral training of three years from the University of Texas at Arlington, USA. Earlier, under the UNDP scheme on capability building he has worked as Associate Professor at the University of Addis Ababa, Ethiopia. He was selected by the BSRB and was appointed as Faculty in the area of behavioural sciences at the Apex training college of Central Bank of India and Dena Bank at Bombay. He has conducted training programmes in many organizations including TELCO, Shipping Corporation of India, Reserve Bank of India, State Bank of India, M.P. Police and Academy of Administration and so forth. He has also done consultancy assignment on capability building of NGDOs for SAP, Nepal in Kathmandu.

Dr. Nagar has extensively published research work on environmental and social issues in reputed foreign/national journals and edited books published by International publishers like Academic Press, Beverly-Hills, Sage, John Wiley and Erlbaum Associates, USA. He has written a book on Human Reactions to Crowding.

ENVIRONMENTAL PSYCHOLOGY

Dinesh Nagar
Director
M.P. Bhoj (Open) University,
Bhopal

CONCEPT PUBLISHING COMPANY, NEW DELHI-110059

Cataloging in Publication Data—DK
***Courtesy:* D.K. Agencies (P) Ltd. <docinfo@dkagencies.com>**

Nagar, Dinesh, 1958-
Environmental psychology / Dinesh Nagar.
p. cm.
Originally presented as the author's thesis (Ph. D.)—Allahabad University.
Includes bibliographical references (p.)
Includes index.
ISBN 8180692663

1. Environmental psychology. I. Title.

DDC 155.9 22

ISBN 81-8069-267-1 (PB)

First Published 2006

Published and Printed by
Ashok Kumar Mittal
Concept Publishing Company
A/15-16, Commercial Block, Mohan Garden,
New Delhi-110 059

Phones : 25351460, 25351794
Fax : 091-11-25357103
Email : publishing@conceptpub.com

Dedicated
To
The Memory of
My Father
Late Dr. Prabhu Shanker Nagar

FOREWORD

For the last three decades or so there has been an increased realization of the disturbed man-environment relationship. The alarming increase in the level of environmental pollution, reduction in the forest cover, ozone depletion and growing tendency of humans to subjugate the natural environment is profusely acknowledged by researchers, social activists, policy makers and even the common man. A number of studies have empirically demonstrated that various social, psychological and health related problems are linked with the fast deteriorating environmental conditions. The above mentioned issues having adversely affected man-environment relationship seems to be one of the major reasons responsible for the emergence of environmental psychology as an independent field of scientific inquiry. It is surprising that most of the existing voluminous texts on environmental psychology available in edited volumes and books are written keeping the eco-contextual reality of Europe and America in mind. The studies done by Indian scholars are not adequately addressed in these volumes. It is only in the recent years that the research contributions made by Indian researchers in some of the major areas of environmental psychology are compiled and presented in edited books. However, so far no textbook is available that covers major issues on environmental psychology and integrates studies done by Indian researchers. These issues are to be transferred to the younger generation through classroom teaching and as such there is no book dealing with environmental issues.

The book attempts to provide the students with most of the relevant information on environmental psychology prescribed in the syllabus. The author makes a comprehensive attempt to generate interest in the readers by providing examples on various environmental issues that are context relevant and friendly. Moreover, the studies conducted by Indian researchers are integrated in various chapters of this book. The book may prove useful in sensitizing the students and researchers who are interested in understanding various intricate issues related to the impact of physical and social features of the physical environment on various measures including social, psychological and health related outcomes.

The book is neatly organized around three distinctive parts incorporating eight chapters that are aimed at discussing studies on major areas of environmental psychology. First section of the book discusses the historical, theoretical and methodological perspective of environment and behaviour. Second section includes the processes involved in environment and behaviour transactions. Finally, the application of the knowledge gained from environment and behaviour studies are discussed comprehensively in the third section of the book.

I hope that the book written by Dr. Nagar will ignite the interest and attract the attention of students and researchers in the exciting areas of environmental psychology and will be of use in environmental planning and management.

(R.S. Sirohi)
Vice Chancellor

Barkatullah University
Bhopal

PREFACE

Earlier the term environment in psychology was used in a restricted sense. Later, the realization of disturbed man-environment relationship and its adverse consequence on human health and well-being were considered one of the potent reasons for the emergence of environmental psychology. Environmental psychology as an independent discipline emerged after 1960s. A number of pertinent environmental issues and human responses to various problems resulting from environment-induced problems were delineated and addressed by researchers mostly working in Europe and America. Influenced by the Western research on environment and behaviour, in the last two decades, the Indian scientists have increasingly paid greater attention to various aspects of environment and behaviour. A number of seminars were organized around major environmental issues in many Indian universities and colleges. In many cases, the contribution made by Indian researchers were compiled in edited books. These endeavours facilitated greater understanding of the complexity of environment-behaviour relationship in the Indian context. However, the existing edited volumes were largely centred on some specific areas of environmental psychology and were largely research oriented. These valuable contributions spelled out the viability of environmental psychology, however, they were not able to adequately stimulate the interest of students whose one of the primary objectives was to get the needed information prescribed in the syllabus in one place. Thus, paucity of the relevant study materials on environmental psychology provided the author an academic stimulus to write a textbook wherein most of the syllabi prescribed at the U.G. and P.G. level courses is covered. While, sufficient information regarding various aspects of environment and behaviour is available in the edited books and textbooks written by foreign scholars. The information provided in these edited volumes is comprehensive and provides useful source of information on various intriguing issues pertaining to environment and behaviour. However, they are mostly centred keeping in mind the Euro-American context. Furthermore, the examples given in these volumes are largely context specific. Moreover, these volumes are in short supply and are not readily accessible to most of the students. The reasons mentioned above together with the interest of the author in the field of specialization were considered convincing to draft a textbook on environmental psychology. In this book attempt has been made to provoke the interest of the prospective readers in the area of environmental psychology by giving examples that are context friendly and are grounded in the Indian cultural reality. Moreover, studies done by the Indian researchers and some of the potent environmental issues are addressed in a capsule form in the present textbook.

The book is organized around eight chapters which may enable the readers to have an overview of the work done by researchers in the major areas of environmental psychology. These chapters are broadly organized around three broad parts. Briefly, part one incorporates two chapters. The chapter one traces historical perspective and scope of environmental psychology while the chapter two provides the reader with the methodological and theoretical perspective of the field. A major focus of the four chapters in part two is on the "processes of Person-Environment transactions". The chapters on basic psychological processes deal with issues pertaining to human cognition and perception, spatial behaviour, crowding and environmental stress. These chapters examine how the abovementioned psychological processes relate to people's everyday transactions with the large-scale environment. Finally, two chapters are subsumed under part three deals with the applications of environmental psychology to community problems.

The first chapter of the book attempts to outline the origin and roots of environmental psychology. A brief discussion about man-environment relationship both from the Western and Indian perspectives are included. Furthermore, definition and conceptualization of environmental psychology, some salient prominent features and trends of environmental psychology are incorporated in this chapter. Finally, a brief focus of Indian researches, future directions and challenges are discussed in the chapter.

Chapter two is devoted to address the theoretical perspectives of environmental psychology. This chapter advocates the utility of scientific methodology for obtaining true knowledge of various aspects of man-environment relationship. Granted the fact that the field of environmental psychology lacks focused direction and is deprived of sufficient data sets, thus, comprehensive theories are not developed that meets the criterion of prediction, summarization and generalization. Nevertheless, dominant theories and models whose predictive validity is limited to specific situations are discussed.

In the chapter three concepts, nature and characteristics of environmental perception and cognition are discussed. Attempt has been made to outline the phenomena related to adaptation, perception of change and perception of natural hazards. Furthermore, issues pertaining to environmental assessment are explored and broad approaches for assessing the scenic quality are addressed. A brief mention on environmental personality is made wherein key elements depicting individual broad disposition towards the environment are discussed. In the preceding section issues pertaining to environmental cognition, cognitive mapping and cognitive maps are discussed and relevant literature are reviewed. The chapter also incorporates various perspectives and theoretical orientation of environmental cognition research. In the concluding section recent trends in environmental cognition research is focused.

In chapter four attempts have been made to explore issues pertaining to human spatial behaviour. A number of salient conceptual and methodological frameworks on personal space and territorial behaviour are presented in this chapter. This chapter also highlights theoretical perspectives on personal space and reviews the applications of personal space research with supportive evidence. Finally, studies on territorial behaviour and its implications in understanding the complexity of environment and behaviour relationship are discussed.

In chapter five attempt has been made to present a brief historical overview of crowding research. Furthermore, three broad categories of research conducted in animals and human subjects both in experimental and field settings are comprehensively reviewed. A separate section is devoted to review studies on crowding done in the Indian context. Finally, the theoretical perspectives and major models of crowding are presented with supportive empirical evidence.

In chapter sixth the concept, major types, characteristics and sources of environmental stresses are discussed. In the preceding section attempt has been made briefly to review studies on major environmental stresses that impinge on us in our daily lives and disturbs us mentally, psychologically and are detrimental to health and well-being. In this regard the effects of noise, air and chemical pollution and their consequences are discussed with suitable examples.

The chapter seventh attempts to examine architectural and design features of the built environment on social behaviour of residents. The chapter attempts to summarize three different viewpoints regarding the degree and extent of architectural influence on human behaviour. Furthermore, a brief mention is made to appraise the readers about the design process and its various stages and how designs are planned and implemented to satisfy the users need. The preceding section review studies on various design features in the residential environment that tends to promote or inhibit social contact and its associated consequences on social and adjustment measures. A number of studies conducted in college dormitories and high-rise buildings are discussed and the role of design features on social behaviour and residential satisfaction are discussed. The chapter attempts to examine reasons as to why some housing projects dramatically fail?

What aspects of the residential design are more vulnerable to crimes and anti-social tendencies? How the physical aspects of the design can be made more users friendly? In the Indian context a case study of Chandigarh city is presented so as to provide the students with a perspective on how the lack of congruence between users need and designers intention hurts the feelings of residents.

Granted the fact that human beings have disturbed the delicate man-environment relationship. The quality of water, air and land is disturbed and is causing hardship to mankind. Within this backdrop the chapter eight attempts to examine how behavioural intervention strategies are helpful in promoting environmentally protective behaviour and discouraging environmentally destructive tendencies. This chapter also attempts to discuss the application of some specific approaches to deal with environmental problems including garbage, littering problems and energy conservation. In the preceding section a number of movements initiated by local community members are discussed wherein evidence of protection of surrounding ecology is highlighted. Finally, two competing theoretical perspectives—Deep ecology and Gaia framework—are presented to provide the reader with academic stimulation about environmental problems and related issues.

The book deals with a multidisciplinary and overlapping subject matter of environmental psychology. This volume will be helpful for the students to appreciate the exciting research that can be conducted in the area of environmental psychology that has both scientific vitality and practical utility.

Dinesh Nagar

ACKNOWLEDGEMENTS

I take this opportunity to express my gratitude to Professor Janak Pandey, Head, Department of Psychology, Allahabad University under whose able guidance I had the opportunity to do my research thesis in one of the major areas of environmental psychology. During drafting this book Dr. Pandey provided me a lot of academic stimulation and helped me in organizing the chapters in three distinctive parts. My interest in the field of environmental psychology was further intensified when I got the opportunity to work with Professor Paul B. Paulus, Head of the Psychology Department at the University of Texas, Arlington, USA, in a longitudinal project on army families living in radically different environmental settings. With Prof. Paulus I got the distinctive opportunity to collaborate in many writing projects in various areas of environment and behaviour. I express my profound thanks to him for helping me in crystallizing my thoughts and concepts on various facets and intriguing issues of environmental psychology.

While working on this volume I have received considerable academic support from Professor Uday Jain, Former Vice Chancellor of A.P.S. University, Rewa, and former Head of the Psychology Department, Barkatullah University, Bhopal. Professor Jain has critically read many chapters of the present volume and provided me valuable comments and suggestions for improvement. I gratefully acknowledge his encouraging remarks. During the work many colleagues and students have helped me in numerous ways. I express my sincere thanks to all of them. I am particularly thankful to Dr. Bhupinder Singer, Reader, Department of Psychology, Barkatullah University, Bhopal and Dr. Medhavi Chourey for comprehensively and meticulously reading earlier drafts of the book. I express my thanks to all my family members for supporting this project.

I am immensely thankful to Dr. Ram Prasad, Former Vice-Chancellor, Barkatullah University, and Dr. Kamlakar Singh, Vice-Chancellor, M.P. Bhoj (Open) University, Bhopal for taking keen interest in the present project and providing me insightful advice from time to time.

I hope that the present book will receive the attention of social scientists, students and professionals who are involved in policy planning and its implementation and will add to the broad area of environmental planning and management. I hope this book will sensitize the students and researchers to pursue their interest in the vibrant areas of environmental psychology.

Dinesh Nagar

CONTENTS

PART 1

CONCEPTUAL AND THEORETICAL PERSPECTIVES OF ENVIRONMENTAL PSYCHOLOGY

1

HISTORICAL PREVIEW OF ENVIRONMENTAL PSYCHOLOGY

Environmental psychology emerged during the 1960s as the result of various technological and societal concerns. People who, a few years ago were only vaguely concerned about the deteriorating quality of their physical environment, have now become increasingly aware and more concerned about the range of environmental issue which directly or indirectly affect them. For instance, emission of noxious gases in the atmosphere from industries and automobiles and dumping of industrial wastes in the water are some of the pressing environmental concerns. Forest cover is gradually reducing because of indiscriminate cutting of trees. The natural habitat is adversely affected and radioactive radiations are slowly penetrating the spaceship earth. Evidence of nuclear accidents is observed in all corners of the world. Floods, famine, earthquakes and changes in atmospheric temperatures are evident. Within this backdrop the world-wide quest to protect the surrounding ecology is intensified not only by environmental activists and social workers but also by the scholars working in diverse areas like geography, psychology, town planning, architecture, forestry, sociology and biology are constantly addressing environmental issues affecting mankind so as to maintain and restore the harmonious balance between man and the nature. In this chapter an attempt will be made to trace the historical development of environmental psychology and to highlight the major areas of research in which the role of physical environment and its impact on human behaviour is examined.

Origin and Roots of Environmental Psychology

Environmental psychology has its origin in biological sciences like zoology and botany. The ecological notions documented around 1900 by biologists examined the correspondence between environmental features and animal species. In the 1920s a group of sociologists adopted the ecological guidelines to study the human population residing in cities. The prevalent methodology used both by the biologists and sociologists interested in ecological principles were largely unobtrusive in nature. These scholars carefully observed and recorded the flow of events to examine the relationship between organism and environment.

Before 1940s the predominant orientation of psychology was looking within the person rather than what was around as a potential source of explanation for behaviour. Most of the times, the individual was treated as if it were was isolated from environmental surroundings. It was during 1944 an ingenious paper entitled "Psychology Ecology" got published by one of the most noted and influential psychologist Kurt Lewin. The focus of this paper implicitly and forcefully acknowledged that the opportunity and constraint of the external environment forces should be carefully examined to understand the behaviour of the individuals. The field of social psychology was further expanded after the pioneering work of Lewin (1951) in which he proposed the formula B = f (P, E) for defining behaviour. The interacting relationship of characteristics of different kinds of people and qualities of environmental contexts/settings was emphasized fully as a determinant of behaviour.

The effects of physical environment on perception and illusions were examined by psychologists (Rivers, 1905). The effects of physical environment on work efficiency, production and human relations were investigated by industrial psychologists (Beutell, 1934). Furthermore, psychologists adhering to behavioural school of thought conceived physical environment in terms of array of stimuli and behaviour as a chain of responses (Skinner, 1953). These studies were largely guided by the presumption that environmental forces shape the behaviour of the passive organism and empirical approaches are helpful in formulating laws of behaviour.

Although, emphasis on the external determinants of behaviour were focused by Lewin and his associates (Brunswik, 1947; Lewin, 1951) social psychology's conception of the environment was narrowly phrased in terms of social environment. For the most part, the ecological context and the physical environment in which the behaviour was embedded were largely de-emphasized. During this period Barker and Wright (1949) published an article outlining both practical and scientific value of ecological viewpoints. They pointed some important limitations of the laboratory experiments and advocated for the development of psychological habitat maps. They argued that the maps would prove useful in documentation of social/psychological environment that surround people and cumulatively affect as well as determine the behaviour, feelings and perceptions. Finally, they pointed out that emphasis on the naturalistic records of behaviour over the existing methods of social psychology would prove helpful for the formulation of general theories about environment and behavioural studies.

The scattered research done largely by some social psychologists during 1940s and 50s emphasized the role of external environmental forces not only expanded the scope of psychology but it also proved helpful in defining the future viability of the field of environmental psychology. During the 1960 the wealth of knowledge led to the convergence of important trends with an expanded methodological perspective. A shift in the focus was visible in the unit of study from single behaviour to complex behavioural patterns, an interdisciplinary perspective, a 'systems' orientation and associated time-linked processes and a more open approach to action oriented problem focused research. In this chapter the historical overview of the environmental psychology will be presented by delineating the salient predominant trends and some supportive studies. Attempt will be made to define the field of environmental psychology and to briefly spell out man's relationship with nature. A brief mention will be made to illustrate the Indian perspective on environment and behaviour. Finally, the present status of environmental psychology will be covered in the concluding section of this chapter.

Salient Predominant Trends

In a capsule form a brief presentation is made regarding predominant trends that facilitated the rapid growth an expansion of the emerging and independent field of environmental psychology.

The Animal and Ethological Tradition

In a controlled experimental setting an initial attempt was made by Calhoun (1962) to examine the impact of crowded environment on various forms of social behaviour in a colony comprising of rats. The phenomenon of 'behavioural sink' was observed as the colony of rats became crowded. The results indicated that population density caused various forms of deviant behaviour including evidence of enlarged adrenal glands which is apparent sign of stress. The results regarding crowding and its associated ill effects were found consistent by other scholars working on animal population (Christian, Flyger & Davis, 1960). The animal ethological tradition also made valuable contribution through its emphasis on naturalistic observation and careful examination of the relationship between organism and the physical environment (Lorenz, 1970; Esser, 1974; Ardey, 1970). The research on non-human subjects stimulated environmental psychologists into action and human studies on environment and behaviour became the focus of attention so as to test the generalizability of results from animals to human population.

Research on Personal Space

The book of Hall (1966) entitled *The Hidden Dimension* ushered a prominent line of research emphasizing the importance of personal space. The research of Hall focuses on four distinct spatial zones, which varies as a function of social context as well as interpersonal relationship. These are labelled as intimate, social, personal and public zones. The cultural difference in personal spacing is also reflected extensively in the work of other scholars (Sommer, 1969). In his book entitled *Personal Space: The Behavioural Basis of Design,* he illustrates the interpersonal and social qualities of the environment in naturalistic settings like hospital, libraries, airport and homes. One significant implication of Sommer's (1969) study reveals the fact that re-arrangement of furniture in semicircular position facilitates the interaction patterns of elderly residents. Another significant contribution of Sommer is his intensive empirical examination of airports where seating arrangements are in straight rows. The passenger reflected greater dissatisfaction while waiting for the delayed flight. Sommer (1974) contended that the sociofugal arrangement of chairs inhibited social interaction among passengers and induced in them the feeling of alienation and dissatisfaction. The implication of personal space studies emphasizing the need for the appropriate interior arrangement of furniture is helpful in generating positive emotional states in the users.

Studies on Behavioural Settings

The research conducted by Roger Barker (1968) was influenced by the notion of life space given by Lewin. Barker recognized the importance of studying behaviour in the environmental context and emphasized on the analysis of "behavioural settings" and taxonomized it along different dimensions. All dimensions of the physical environment, which has psychological import, were included in behavioural settings. The work of Barker and his associates deal with the physical/ social situations in which behaviour occurs. A wide range of behavioural settings like school, the playground, machine shops, and supermarkets were extensively observed and described. Their ecological viewpoint acknowledges that people are one component of the behavioural setting system. The behavioural settings are not passive but are self-regulating where range of behaviours and actions of people are promoted or restricted (Barker & Wright, 1949; Barker, 1978). The research of these scholars emphasizes unobtrusive observation of complex behavioural patterns in naturalistic settings.

The Growth of Environmental Psychology

It was in 1976 the second edition of the first textbook of Proshansky, Ittelson and Rivlin entitled *Environmental Psychology: Man and his Physical Environment* appeared. Many disillusioned social and traditional psychologists turned their attention to the study of large-scale physical environment and its relationship to social processes and range of behaviours. A focus of molar, global unit of analysis, social problem centred research with an interdisciplinary ethos and more emphasis on naturalistic research guided the rapid growth of environmental psychology as an independent field of scientific inquiry.

Partnership between Professionals and Psychologists

The equal status partnership between architectural designers and environmental scientist lead to the establishment of a common forum called the Environmental Design Research Association (EDRA) in the late 1960s. An interdisciplinary effort to understand the congruence between design features and users needs was noted by environmental scientists. The traditional designers sole emphasis on aesthetics of the built environment, disregarding the inhabitants preferences, needs and social behaviour was subjected to criticism. The design features and the appropriate use of space inhibit or promote certain forms of desirable or undesirable behavioural pattern were emphasized (Yankey, 1972; Newman, 1972).

Publications of Journals and Books

Research on environment and behaviour spurred during 1960s. It was of the short gun quality. The earlier enthusiasm of environmental psychologists slowly became more integrated, channelized and focused across time. The appearance of a research journal entitled *Environment and Behavior* in 1969 marked the first scholarly outlet for researchers. The scattered research of the 60s and 70s was nicely integrated in edited volumes, monographs and textbooks (Ittelson, Proshansky, Rivlin & Winkel, 1974; Stokols, 1972; Freedman, 1975; Down & Stea, 1977). For the first time "The Annual Review of Psychology" had incorporated the survey of emerging field dealing with environment and behaviour (Craik, 1973). At periodic intervals chapters dealing with the research focusing on environmental psychology had appeared (Stokols, 1978; Russel & Ward, 1982; Holahan, 1986). During the first two and half decades systematic research work had been undertaken by scholars on topics such as personal space, crowding, territoriality, social and environmental impact assessment, post occupancy evaluation of residential environment and environmental impact in offices and factories. This rapid expansion of knowledge spurred the establishment of numerous research journals (*Population and Environment,* 1978; *The Journal of Environmental Psychology,* 1981; *Journal of Architectural Planning and Research,* 1984; *Journal of Complex Environmental issues,* 1989) and numerous books and handbooks (Paulus, 1987; Jain, 1987; Fisher, Bell, & Baum 1984; Evans & Cohen, 1987; Stokols & Altman, 1987). Indications of quantitative growth of knowledge on environment and behaviour have yielded sufficient evidence of the practical utility and future viability of environmental psychology. This assertion is reflected in an earlier review of (Stokols, 1978) stating that "Environmental psychology is no fad of 1960s— a short-lived product of environmentalists and political activists. Instead, this field has taken hold both conceptually and empirically. It is now comprised of several active and focused research domains. The scientific vitality of environmental psychology is reflected in the substantial theoretical and empirical progress that has been made within many of its major sub-areas."

Man-Environment Relationship

Keeping the basic nature and distinctive features of human being and environment some pertinent questions require serious debate and discussion. For instance, it is pertinent to examine the issues regarding the quality of prevailing relationship between man and nature. What type of relationship exists between man and nature? Is the relationship between man and nature harmonious enough to sustain each other? If not, how to restore the mutually beneficial and healthy relationship between man and the nature?

There are three prevalent views regarding relationship between man and the environment (Wohlwill, 1973). These viewpoints range from person dominated and environmental determinism to interactive and dynamic interplay between man and his surrounding environmental features. The *environmental determinism* implies that external forces dictate the range of responses and require a person to behave in a specific way. Much of the earlier studies conducted by environmental psychologists point out that the physical environment can elicit positive or negative attitudes and feelings. This viewpoint downplays the ability of the individuals who self select the environment so that their needs and aspirations are fulfilled. The second kind of environment-person relationship can be described as *possibilism*. Certain qualities of the physical environment are associated with approach or avoidance behaviour. This aspect of behaviour/environment relationship may explain a variety of reported differences between urban and rural incidences, psychosomatic disorders as well as diffusion of responsibility and bystanders apathy. The third kind of relationship is one in which the environmental features serves as a motivating force. The important issue concerning behavioural and psychological adaptive capabilities of people to a wide range of environment may be explained by applying this environment/behaviour framework. The three important facets encompassed in this form of relationship are the following: (1) affective and attitudinal responses to environmental features, (2) approach and avoidance response to various characteristics of the environment and (3) adaptive processes to environmental attributes. Much of the research on environment and behaviour has discussed the adaptation process in response to features of the physical environment like noise, pollution, temperature and ghetto life.

Man-environmental interaction from the human point of view can be subsumed under two major categories. In the adjustive-accommodative relationship the available environmental resources are used to satisfy the survival needs of the individual. In this type of relationship individual is not attempting to subjugate nature, instead, to live within the regularities of nature. For instance, houses are constructed in those areas that are beyond the reach of floodwater. Thus in this type of person/environment interactions individual is dependent on nature with maximum amount of accommodation and minimum amount of manipulation. In the manipulative-exploitative interaction there is maximal human effort for subjugating and conquering the nature to satisfy the human wants and needs. The focus is on maximum gains in favour of individual and disregarding the injury that has been caused to nature. The dumping of industrial wastes and garbage, air pollution are some examples of such kind of relationship.

The fourfold classification focusing on the relationship between person and environment is comprehensively discussed by Altman and Rogoff (1987) in the "Handbook of Environmental Psychology". The attempt has been made to integrate a number of researches conducted in the field of environmental psychology. The fourfold worldview deals with the trait, interaction, organism and transactional perspectives. A brief mention of these perspectives are given below:

The *trait approach* focuses on individuals and de-emphasizes the role of environmental features and contexts. This approach has received limited attention from environmental psychologists.

The *interactive perspectives* are prevalent in contemporary environmental psychology and require elaborate attention. The dominant focus of interactional approach is on the relationships between environmental features, which are usually treated as predictors, and behavioural and psychological variables as outcomes. Typically, the unidirectional or linear theoretical models are tested to study the relationship between environment and behaviour. Finally, temporal factors are used to locate and describe the changes in psychological functioning. The straightforward unidirectional effects of crowding and psychological functioning are examined within the interactional framework (Baum & Paulus, 1987). The interactional perspective is highlighted in many researches, which deals with the environmental influences that facilitates or inhibits the regulation of group interaction in a variety of contexts (Paulus & Nagar, 1987). Furthermore, studies on defensible spaces, housing, and workplaces have also adopted the interactional perspective to explain how person and group factors interact with environmental features to affect satisfaction, performance and outcomes (Altman & Rodoff, 1987). Thus, in order to explain the causal links between environment and behaviour, the interaction approach treat environmental contexts, psychological processes and temporal factors as separate entities.

The *organism approaches* take as a unit of study the qualities of a holistic system and emphasize on reciprocal and complex pattern of relationship between various elements of the integrated system. The tendency of organism is to regulate the operation of the system and to achieve homeostatic. The organismic viewpoints are not widely used in environmental psychology. However, in recent years this approach has been used to explain complex relationships between environment and behaviour. Several models of crowding have adopted holistic frameworks (Sundstrom, 1978; Altman, 1975). Crowding is conceived, as a complex system comprised of various personal, interpersonal and physical components, which exert reciprocal influence to one another thereby affecting the underlying processes like psychological appraisal of the situation and stress. Adopting a holistic system framework Moos and Lemke (1984) have studied various settings like college dormitories, schools and hospitals. He envisions these settings as comprising of environmental and personal subsystems, which mutually influence one another from multifaceted paths and directions. Similarly, Stokols and Navoco (1981) developed a holistic model to analyze the relation between psychological aspects of transportation and well-being. The central theme of the organismic perspective is its emphasis on the holistic psychological systems composed of interrelated person and environment components, which potentially exert mutual and reciprocal influence.

The transactional approach to study the environment-behaviour relationship is advocated by a number of scholars largely because the field of environmental psychology stresses on molar physical aspect of the environment (Proshansky *et al.*, 1976; Ittelson, 1973). For instance, Ittelson (1973) has accepted the utility of the transactional idea and stated that "man is never completely encountered independent of the situation through which he acts, nor is the environment ever encountered independent of the encountering individual. It is meaningless to speak either as existing apart from the situation in which he is encountered" (p. 19). Recently, many environmental psychologists have emphasized the need to deal with the man-environment transactions both empirically and theoretically (Stokols & Shumaker, 1981). However, these ideas, which are accepted in principle and not translated into action (Altman & Rogoff, 1987). The ecological research conducted by Barker (1978) and Wicker (1979) in a range of environmental settings like schools, small towns and grocery stores across a period of time closely corresponds with transactional notions. These scholars believe that patterns of behaviour are inextricably linked with the physical and social environment and can be understood only when viewed in a continuous and elaborate flow and temporal context. Wicker (1987) used the concept of behavioural settings and setting programme to explain and describe the dynamic qualities of the person-environment relationship. The effects of undermanning and overmanning in school setup and attempt to maintain the ongoing school programmes like debates, sports are comprehensively addressed within the transactional framework. The transactional model provides a framework in which the individual is seen as the active recipient and can influence the environment by personal actions. For instance, to achieve the best fit between user's need and environment design features, inhabitants may change the physical environment (e.g. fencing the area around the building, encroachment), which is congruent with their needs and aspirations (Rapoport, 1985; Michelson, 1977). With supportive data Paulus and Nagar (1987) highlights the viability of the transactional model to explain the environmental influences on social cohesion and group development. The dialectic analysis of privacy regulation as a holistic multi-mechanism process also illustrates the applicability of transactional approaches (Altman & Chemers, 1980). To summarize, the transactional approach deal with various salient orientation of the field and thus have broad appeal to environmental psychologists. These approaches treat the integrity of person, psychological processes and environmental context not as a separate element but as a mutually interlocked aspect of the whole. The temporal dimensions are intrinsically linked with the whole to describe and understand the stability and change in the phenomenon.

The Construct of Environment

The construct of environment is profusely used both in biological and social sciences. Each discipline defines it according to its unique requirements and theoretical perspectives. Environment as a construct is multifaceted and has been extensively used by contemporary scholars in psychology. However, as has been emphasized by Tripathi (1997) "Whether it is extreme environmentalism of Watson, or the concept of 'press' in Murray's personology or the behaviour considered as a joint function of person and environment in Lewin's dynamic psychology, interactional features of environment has not been incorporated adequately in its conceptualization." However, it is only after the emergence of environmental psychology the conceptualization of environment both as an entity in determining human behaviour as well as being continuously modified by human behaviour was emphasized.

A variety of terms has been used to describe the person-environment relationship. The physical environment connotes every aspect of the ecological reality that surrounds a person. This conceptualization of physical environment is very comprehensive and broad. Usually, environmental psychologists distinguish the physical environment into two types—the built or designed environment and the natural environment. A small scale aspect of the built environment may focus on the rooms of the buildings, arrangement of furniture in the room. On a larger scale, one can focus on the layout of the range of residential, hospital, prison, offices and factory buildings in which behaviour occurs. Finally, the built environment at the even larger scale encompasses the neighbourhood, the city as well as entire region. Various aspects of the built environment have generated considerable attention of environmental psychologists because the built features have resulted in promoting or inhibiting socially desirable or undesirable behavioural patterns.

Those aspects are subsumed under the natural environment where human interventions to manipulate or modify the setting are at least minimum or not present. The national parks, recreational and wilderness areas are some examples of natural environment where studies are conducted to examine the relationship between perceived natural environment and users behaviour. Issues associated with the natural environment may also deal with natural hazards like earthquakes, floods, social impact assessment and so forth. Researchers have frequently used various observational techniques and unobtrusive measures to study the behaviour of children in schools, playgrounds or classroom settings.

The environmental psychologists also recognize the role of social and interpersonal factors that influence the response of individuals to the physical environment. Researches on social factors that are not separated from the physical features of the environmental settings like privacy, personal space and crowding receive wide coverage in the literature to understand the environment behaviour links.

Towards a Definition of Environmental Psychology

One common practice to introduce any field of scientific inquiry is to provide some popular definitions because it provides useful summaries of the basic principles of the field. However, definitions are narrow in scope and it sets rigid boundaries with over-simplified summaries of the field. Besides, definitions become soon outdated because the field is always growing and changing with the production of new researches. The field of environmental psychology has grown considerably but as far as the theoretical formulations are concerned it is still very immature. One strategy to define the field is to closely examine the research activities. Despite the diverse areas of interest, one common concern of researchers is examining the relationship between man and the physical environment. Although, rephrased differently most of the scholars agree that the discipline is concerned with the relationship between physical environment and human behaviour. For instance, Burroughs (1989) defines environmental psychology as "the study of the interrelationships between physical environment and human behaviour." In a similar vein, Gifford (1987) defines environmental psychology in terms of the transactions between individuals and their physical settings. The reciprocal relationship between man and the environment is emphasized in both the definitions. The person-in-context approach is the central theme in environmental psychology which advocates that for understanding of behaviour we need to look independently not only within the person and the surrounding but also take into account the interaction between person and the environment. In other words, the interactional perspective is rooted as central principle of environmental psychology.

The field of environmental psychology was initially called "architectural psychology" (Cantor, 1970) because the roots of the field were primarily focused on physical features of design of the building and its implications on human behaviour

(Osmond, 1957). Thus, the exclusive focus on the physical environment was understandable. However, research focus on range of topics like crowding, personal space, and urbanization expanded the scope of the field with both physical and social features of the environmental setting/context included in defining environmental psychology. As has been pointed out by Proshansky, Ittelson and Rivlin (1976) that "The physical environment that we construct is as much a social phenomenon as a physical one (p. 5)." Similarly, cited in the book of Cassidy (1997), the French psychologist Claude Levy-Leboye concludes, "The physical environment simultaneously symbolizes, makes concrete, and conditions the social environment (p. 15). There is currently general agreement among scholars that environmental psychology is the study of the transactions between individuals and their socio-physical environments (Bonnes & Secchiaroli, 1995; Altman, 1976; Stokols, 1978; Cassidy, 1997).

Salient Features of Environmental Psychology

An alternative way in which we might define environmental psychology is to discuss some of the major orientation, unique values and basic principles of the field. Furthermore, how the scholars working in the area of environmental psychology define their problems, use appropriate methodology and examine the phenomena of interest is briefly subsumed below:

A Holistic Perspective

The approach of environmental psychology as viewed by several proponents is dynamic and holistic rather than reductionist (Ittelson *et al.* 1974; Altman, 1974). A reductionist approach is analogous to trying to complete a jigsaw from the individual pieces with no idea of what the complete picture looks like (Cassidy, 1997). In contrast, the holistic approach emphasizes on complex pattern of reciprocal relationship between person and the environment components. An implicit principle underlying the global and molar perspective is that the researcher is dealing with person-environment units in which unidirectional links between various components comprising the whole is examined. The focus is on complex causal relationship with environment and behaviour. The studies of Barker (1968) and Wicker (1979) in a range of settings like churches, schools, playgrounds are examples of the application of holistic perspective which includes the contributions of physical, social and psychological components. The holistic perspectives are accepted in principles by most of the scholars working in the field of environmental psychology (Altman & Rogoff, 1987; Ittelson, 1974).

An Interactional Perspective

As advocated by Lewin (1951) behaviour is a function of the person, the environment and the interaction between the two is accepted in environmental psychology.

A Model of Person as Active rather than Passive is Emphasized

The person in environmental psychology is given some degree of freedom and autonomy rather than treating the person within the deterministic framework.

A Social Problem Centred Orientation

One of the most distinctive and desirable features of environmental psychology is rooted in its problem-centred orientation rather than theory centred set of activities. The research sets its goal organized around solution of social problems that concern the people in general directly. Studies dealing with practical social problems have come from professional design field like architecture and urban planning and environmental practitioners. Thus, one major thrust of environmental research is to translate research knowledge into practically useful information that has the potential application. Evidence exists regarding the applications of environmental behaviour research to various socially relevant issues like residential design, community problems and planning and urban stress.

An Interdisciplinary Orientation

The contemporary research on environment and behaviour is inherently interdisciplinary in scope, and encompasses a spectrum of professional inputs from a range of discipline like architecture, urban and regional planning, geography, sociology, organizational behaviour and so forth. Owing to the complexity of socially relevant problems, collaboration of psychologists and professional are increasing. Synergy of ideas from an array of disciplines to address the community problems offers a healthy perspective and new theoretical possibilities to the field of environmental psychology.

Methodological Eclecticism

The attention of environmental psychologist is focused around the question concerning "what is it we study" and what we believe to be the appropriate causal inquiry model. Owing to the complexity of the field and interdisciplinary ethos, one seeks a pragmatic methodological eclecticism with research design and problems guiding the methods. Environmental psychologists are receptive to using a wide range of research methods. Both qualitative and quantitative methods are used were basic and applied research aims to complement each other. Many scholars have broadened their methodological perspective by adopting a 'social unit' and 'social system' inquiry model. They have generally accepted both experimental and field observational techniques. Of particular potential relevance to environmental psychology is methodological theorizing of Campbell and Stanley (1963) which focuses on the utility of quasi-experimental designs to assess the impact of environmental changes created by the intervention of architects in various environmental contexts like high-rise residential accommodations, dormitories and factories. Evidence suggests that certain person-environment relationships are best tapped by naturalistic observation as subscribed by Barker (1968). The different methodological strategies adopted by Tuan (1974) in his historical analysis of cities and community and Rapoport (1980) anthropological analysis of how cultural values interact with environmental resources, technology, and climate affect environmental design are conducive to environmental psychologists. In short, contemporary environmental psychologists value the research problem guiding the appropriate method of study and acceptance of broad methodological perspective is a keynote to contemporary environmental psychology research.

An Emphasis on Naturalistic Research

This is based on the recognition that laboratory experiments are questioned for their artificiality and generalization of results from experimental setting to real life settings. Moreover, experimental studies are low on ecological validity.

An Emphasis on the System Model of Relationship between Environment and Behaviour

The dynamic inter-relationship between aspects of the environment and behaviour is recognized and an awareness of how changes in one part will affect the other part is examined.

An Emphasis on Multiple Levels of Analyses

Both individual as well as societal/organizational levels of analyses is emphasized with a particular focus on molar level.

Traditional Indian Viewpoints on Man-Environmental Relationship

In a recent review of Indian research on environment and behaviour Jain and Palsane (2004) made a brief mention about the environmental issues in relation to human being. Some useful insights of man's symbiotic relationship with nature can be derived from the traditional Indian texts and books on religion and medicine. While the scientific basis regarding the man-environmental relationship is not apparent in the ancient Indian literature, nevertheless evidence of environmental degradation by human intervention is also not documented. It was accepted that one form of life can protect other forms of life without consuming it (Deodikar, 1991). The underlying theme behind such prescriptions may be related to the principle of integrated 'cosmos' in which man has been considered as one element of equal status amongst other elements keeping symbiotic relationships. Furthermore, citation of harmonious relationship between man-environmental based on moral principles can be seen (Jha, 1991; Kulkarni, 1991). The ancient pattern of Indian thinking provides evidence of symbiotic or mutually sustaining man-environmental relationships (Tripathi, 1997). The ancient texts warn man not to disturb the harmony with the nature and to protect the plants and animal species for his own well-being. The Kautilya's *Arthasastra* provides a detailed account of environmental management and prescribes punishments in case of violation. The unique medicinal properties of different plants, herbs and shrubs were known even to common man in ancient times. Even today people living in tribal areas and villages are equipped with such knowledge regarding the curative properties of different plants and vegetation for different ailments.

The ancient writings profusely provide evidence of man's emotional ties with nature. The constituents of universe like *fire, water, air, earth* and *ether* were equated with deities of *agni, varun, pawan, vasundhara* and *indra*. Furthermore, a number of trees were considered as the seat of supernatural beings and were symbolized as having imperishable continuity of life. For instance, the *peepal* tree was respected and worshipped because a deity Vasudeva lives there. *Neem* tree was regarded as the place of Goddess Kali. Similarly, most of the major rivers like Ganga, Yamuna, Saraswati, Narvada were equated with mother like deities. The ancient texts probably induce man to have an emotional and sentimental attachment to various ecological constituents of nature. The appropriate norms and rules were clearly laid down and exploitation of nature was considered sin and was prohibited. Before the construction of houses and ponds, *Yagna* was performed as part of self-enhancing duty. There are many instances in which the natural forces are seen as endowed with supernatural and divine powers. In some pagan worship practices even today in many tribal and rural areas components of natural environment are seen as forces that are respected and feared rather than subjected to exploitation.

Indian Research

Jain and Palsane (2004) presented a brief overview of ICSSR surveys with regard to environment and behaviour studies. The first survey edited by Mitra (1972) incorporates very few studies focusing on environmental-behaviour relationships (A.K.P. Sinha, 1972; D. Sinha, 1972; M.M. Sinha, 1972). In these studies environment was treated as separate entity and unidirectional relationships were explored to predict outcomes. The second ICSSR survey (Pareek, 1981) included four articles under the heading environmental psychology and social issues. The social system perspective was adopted to analyze environmental issues (Padaki & Vyasula, 1981). The review failed to discriminate between social problems and environmental issues. Other chapters of this section (Rao, 1981; Moulik, 1981) failed to adequately address the socio-physical issues. The third ICSSR survey comprising of three volumes (Pandey, 1988) did not include any chapter on environmental psychology. One chapter in the fourth ICSSR survey (Pandey, 2004) is devoted exclusively on environment and behaviour (Jain & Palsane, 2004). This chapter has comprehensively reviewed various studies undertaken by Indian researchers, which directly deal with issues falling in the areas of environmental psychology. Based on the proceedings of various seminars and conferences, edited volumes on environment and behaviour are published (Jain, 1997). In addition some volumes are also available on major areas of environmental psychology (Jain, 1997; Jain, 1987; Nagar, 1998).

Scanning the Indian work in the area of environmental psychology, it seems that earlier studies were more focused on social aspects and the role of social/physical aspects of the environment was not given adequate attention. However, it appears that gradually interest in this field is evolving and more researches are being undertaken on various environmental issues and problems.

Recent Trends and Future Directions in Environmental Psychology

In a recent article Evans (1996) has recapitulated the current trends in environmental psychology. He feels that environmental psychology has lost much of its viability as an independent field within psychology over the past decade. Much of the early enthusiasm in North America has been waned and very little collaborative ideas are forthcoming between behavioural scientists, architects and designers. However, the satisfactory aspect is that research generated in environmental psychology has been fully adopted into mainstream psychology.

"The popularity of environmental psychology can be inferred from various vantage points. For instance, the subscription rates of three major journals in the field, *Environment and Behavior, Journal of Environmental Psychology,* and *the Journal of Architectural Planning and Research* are at all times high. The courses on environmental psychology in North America are attracting students in large numbers. A number of text-books on environmental psychology have published new editions (Baum, Fischer, Baum & Greene; Gifford); two new texts have been published in the US (McAndrew; Veitch & Arkelin) in 1990s. Furthermore, Bonnes and Secchiaroli text is published in Italy (Evans, 1996)." In India Tripathi (1997) has published a text-book on environmental psychology in the native Indian language. In India the undergraduate and post-graduate courses on environmental psychology are offered and are well received by students in major universities and colleges. Books are published in major areas of environmental psychology (Jain, 1987; Nagar, 1998) and articles on environmental psychology are incorporated in many visible volumes on various branches of psychology (Jain and Palsane, 2004; Nagar, 1999; Nagar and Sharma, 2005). However, in the

Indian context adding the prefix 'environment' is sometimes noticed together with a trend of compiling articles presented in seminars and proceedings in edited volumes on various general themes of environmental psychology (Agarwal & Saxena 2003; Jain, 1997). Nevertheless, all these endeavours considered together are index of the popularity and growing visibility of environmental psychology.

The Cambridge and Gower's Ethnoscape series on environment and behaviour have received overwhelming support from readers and each series are having several volumes in the pipeline (Evans, 1996). *The Handbook on Environmental Psychology* sold out of press (Altman & Stokols, 1987) and is now been reissued by Krieger Publications. Evans (1996) in his article reported that the handbooks of both social and health psychology, and new handbook of psychophysiology contains chapters on environmental topics. Cognitive mapping has emerged as an important area in cognitive sciences. Various journals devoted on cognition are incorporating more and more articles on spatial memory, way finding, and computational model of environmental cognition. Furthermore, Environmental education is emerging as a major sub-area within educational curriculum. Research studies in the area of child psychology and lifespan development have started focusing on immediate and background setting and its linkages with healthy development.

The education of architects include some exposure to human behaviour and behavioural science. The idea that features of the physical design affect users is more or less accepted by design professionals. While architects who are in practice have not accepted much of the knowledge of behavioural science. However, in some specialized fields like interior designs and landscape architects have shown considerable interest in many aspects of environmental psychology induced knowledge and have collaborated with behavioural scientists. In many countries the new field of Investigative Psychology is contributing to the criminal justice system. This new field covers major areas of environmental psychology including place theory, territorial behaviour, environmental cognition, prison studies and so forth.

In United States and Europe considerable efforts are evident in understanding life in space and developing programmes for housing travellers and workers in outer space (Evans, 1996). In the field of human factor engineering health and safety issues in the workplace is attracting interest and the inputs from environmental psychology is put into use. Indoor air quality and stress in work place are topics, which are generating renewed interest.

Human response to various spaces prior to their actual development is attracting researchers. The computer simulation techniques are being used and more practical applications of environmental psychology are analyzed. Research on environmental stressors like crowding, noise, pollutants as well as natural and technological disasters are studied and its relationship with health and cognition are examined. During emergency situations the natural behaviour of people have provided valuable insights and leads to emergency planning policies so as to reduce harmful effects. Studies on multiple stressors, cross-situational effects from home to work place and *vice versa* are also attracting researchers' attention.

Another important area is to study global environmental issues. Evans (1996) has reported that the human dimensions of global change has emerged as an important priority area in US and several environmental psychologists like Paul Stern and his associates have published a recent monograph in this area. Besides Annual Review of Psychology, the *Journal of Environmental Psychology* has recently published a special issue on the topic. A number of researchers are focusing their study on the relationship between global environmental changes and psychology. This area derives insights from operant paradigms and motivational theories centred around, explaining how to arrest ecologically destructive behaviours.

In the third world countries people are living in suboptimal conditions. In the third world substantial population is suffering from day to day environmental problems including substandard housing and unhygienic living conditions surrounded with dirty, filthy atmosphere. The focus of research is to connect poverty and poor environmental quality within the framework of environmental psychology. A number of studies are conducted on poverty in the Indian context. However, in most of the studies the role of physical environmental features are either ignored or deemphasized. This gap is realized and some modest attempts are made to study Indian villages taking the physical environmental aspects and examining their relationship with community relationships and health measures (Tripathi, 1993). Furthermore, only meagre attempts and research work on Indian slums are conduced by psychologists (Sandhu, 1987; Nagar & Misra, 2001). For instance, a study was planned to examine the impact of rehabilitation of poor slum inhabitants who were previously exposed to unfavourable environmental conditions. The focus of this study was given on environmental quality, crowding and health (Nagar, 2005). One major result of this study indicated that slum inhabitants who were provided ownership of their slum units by the government exhibited a tendency of improving the environmental quality of their individual and shared dwelling and neighbourhood environment and reported better health relative to their counterparts who were residing in the unauthorized piece of land or were not eligible for ownership of slum units. It seems that feeling of ownership of slum unit is one important factor in protecting and improving the environmental quality of the neighbourhood.

Environmental psychologist should undertake studies on poverty stricken communities and groups who happen to reside in villages and tribal areas within the vicinity of Narvada valley project for generations. How these people respond to forced and imposed dislocation in the name of rehabilitation from the government and what are its likely consequences is an important line of research. Similarly, some researches are conduced in congested inner city areas where community groups, crowding, and environmental quality are connected with social support, tolerance and health measures (Nagar & Sharma, 2005). In addition, more systematic studies which incorporates physical environmental features like type of housing, poor environmental quality, scarcity of resource conditions and poverty dimensions particularly in rural areas, developed-underdeveloped villages, tribal-non-tribal and different ethnic groups are needed to enhance our understanding of environment-behaviour relationships in the Indian context.

Cultural specific studies are also needed to tap the traditional understanding on managing the environmental problems. For example, in traditional Indian texts have sizeable coverage of the importance of physical environments and required attitudes and behaviour on the part of human beings. Modern man has ignored such a rich tradition in order to emulate the Western concept of modern development. The popular paradigm of development necessitates the exploitation of environmental resources without any limit. Such exploitation of physical resources has already created an imbalance in the availability of environmental resources and their requirement for human population on this earth. The fact, that current environmental problems are the result of pursuing the modern paradigm of development. To create a balance between availability of resources and the requirement of a new paradigm for development may be needed. Knowing fully well that the human population at this juncture has reached to state of no return to traditional lifestyle, however, care of the environment is nonetheless most important. Obviously this care needs proper attitudes, proper management of the available resources, proper management of the environmental resources, and eco-friendly behaviour even at the cost of little discomfort. Hence, environmental psychologists are called for exploitation of cultural sentiments and translating them in a scientific language so that it becomes convincing to the people already adapted to a life, which is anti eco-friendly.

The above discussion represents only a sample of current research trends and future direction of research in environmental psychology. It is important to undertake studies across cultures and across different ethnic lines to understand the similarities and differences in responses to various environmental provisions. The moderators and mediator variables may be identified and more precise test may be undertaken to understand the complexity of environment-behaviour relationships.

Summary

In this chapter an attempt is being made to trace the historical development of environmental psychology. A brief presentation is made regarding some predominant trends that facilitated the rapid growth and expansion of environmental psychology as an independent field of scientific inquiry. In this regard major studies conducted on animals are discussed. Furthermore, research on personal space and studies conducted in various behavioural settings are discussed in the capsule form. The growth of the field was marked by prominent publications and establishment of journals is discussed. Finally, inter-disciplinary focus of the field was highlighted by establishment of common forums where design professions and environmental scientists shared their viewpoints.

The chapter addresses issues regarding relationship between man and the environment. What type of relationship exists between man and the nature? Is the relationship between man and the nature harmonious enough to sustain each other? If not how to restore the mutually rewarding relationship between man and the environment? A brief mention is made in the chapter with an attempt to discuss fourfold worldview—the trait, interactive, organismic and transactional perspectives.

The construct of environment is profusely used both in biological and social sciences. Each discipline defines it according to its requirement and theoretical perspectives. The concept of environment as used by environmental psychologist and some popular definitions as given by eminent scholars are discussed. The salient features, major orientation, and unique values of the field are summarized. Finally, traditional Indian viewpoints on man-environment relationship and major research areas undertaken by Indian researchers are highlighted in a capsule form. The chapter ends with a brief note on recent trends and future directions in environmental psychology.

REFERENCES

Agarwal, A., & Saxena, A.K. (2003). *Psychological Perspectives in Environment and Developmental Issues*. New Delhi: Concept.

Altman, I. (1975) *Environment and Social Behavior: Privacy, Personal Space, and Territory and Crowding*. Monterey, California: Brooks Cole.

Altman, I. (1976). Environmental psychology and social psychology. *Bulletin of Social and Personality Psychology*, 2, 96-113.

Altman, I., & Chemers, M. (1980). *Culture and Environment*. Monterey, CA: Brooks/Cole.

Altman, I., & Rogoff, B. (1987). World's view in psychology: Trait, interactional, organismic, transactional perspectives. In D. Stokols, & I. Altman (Eds.), *Handbook of Environmental Psychology* (Vol. 1. pp. 7-40). New York: John Wiley.

Altman, I., & Stokols, D. (1987). *Handbook of Environmental Psychology*. New York: Wiley.

Ardey, R. (1970). *The Social Contract*. New York: Atheneum.

Barker, R.G. (1968). *Ecological Psychology: Concepts and Methods for Studying the Environment and Human Behavior*. Stanford, CA: Stanford University Press

Barker, R.G. (Ed.). (1978). *Habitats, Environments, and Human Behavior: Studies in Ecological Psychology and Eco-Behavioral Science from the Midwest Psychological Field Station*, 1947-1972. San Francisco: Jossey-Bass.

Barker, R.G., & Wright, H.F. (1949). Psychological ecology and the problem of psychosocial development. *Child Development*, 20, 131-143.

Baum, A., & Paulus, P.B. (1987). Crowding. In D. Stokols & I. Altman (Eds.), *Handbook of Environmental Psychology* (pp.533-570). New York: Wiley.

Beutell, A.W. (1934). An analytic basis for a lighting code. *Illuminating Engineering*, 27, 5-16.

Bonnes, M., & Secchiaroli, G. (1995). *Environmental Psychology: A Psycho-Social Introduction*. London: Sage.

Brunswick, E. (1947). *Systematic and Representative Design of Psychological Experiments* .LA: University of California Press.

Burroughs, W.I. (1989). *Applied environmental psychology*. In W.L. Gregory & W.J.

Burroughs (Eds.), *Introduction to Applied Psychology*. London: Scott Forsman and Company.

Calhoun, J.B. (1962). Population density and social pathology. *Scientific American*, 206, 139-148.

Campbell, D.T., & Stanley, J. C. (1963). *Experimental and Quasi-Experimental Designs for Research*. Chicago: Rand McNally.

Cantor, D. (1970). *Architectural Psychology*. London: Royal Institute of British Architects.

Cassidy, J. (1997). *Environmental Psychology: Behavior and Experience in Context.* Sussex: Psychology Press.

Christian, J., Flyger, V., & Davis, D. (1960). Factors in the mass mortality of a herd of Silka deer cervus nippen. *Chespapeake Science,* 1, 19-95.

Craik, K.H. (1973). Environmental psychology. *Annual Review of Psychology,* 24, 403-422.

Deodikar, S.G. (1991). Ecological awareness reflected in Manusmriti. In V.N. Jha (Ed.), Proceedings of the National Seminar on Environmental Awareness, Reflected in *Sanskrit Literature* (pp. 93-100). Poona: Centre of Advance Study in Sanskrit.

Downs, R.M., & Stea, D. (1977). *Maps in Minds: Reflections on Cognitive Mapping.* New York: Harper & Row.

Esser, A.H. (1974). Environment and mental health. *Science, Medicine, and Man,* 1, 181-193.

Evans, G. (1996). *Current Trends in Environmental Psychology.* IAAP Newsletter.

Evans, G., & Cohen, S. (1987). Environmental stress. In D. Stokols and I. Altman (Eds.), *Handbook of Environmental Psychology,* New York: John Wiley & Sons.

Freedman, J. (1975). *Crowding and Behavior.* San Francisco: Freeman.

Fisher, J.D., Bell, P., & Baum, A. (1984). *Environmental Psychology* (2nd ed.). New York: Holt, Rinehart, & Winston.

Gifford, R. (1987). *Environmental Psychology: Principles and Practice.* Boston: Allyn & Bacon.

Hall, E. (1966). *The Hidden Dimension.* New York: Doubleday.

Holahan, C.J. (1986). Environmental psychology. *Annual Review of Psychology,* 37, 381-407.

Ittelson, W.H., Proshansky, H., Rivlin, L.G., & Winkel, G.H. (1974). *An Introduction to Environmental Psychology.* New York: Holt, Rinehart & Winston.

Jain, U. (1987). *The Psychological Consequences of Crowding.* New York: Sage.

Jain, U. (Ed.) (1997). Perspectives on Environment and Behavior. New Delhi: Har-Anand. *Project Report,* New Delhi: ICSSR.

Jain, U., & Palsane, M.N. (2004). Environment and behavior. In J. Pandey (Ed), Psychology in Indian Revisited: Developments in the Discipline (Vol. 3): Applied *Social and Organizational Psychology* (pp. 261-308). New Delhi: Sage.

Jha, V.N. (Ed.) (1991). *Proceedings of the National Seminar on Environmental Awareness Reflected in Sanskrit Literature.* Poona: Centre of Advance Study in Sanskrit.

Kulkarni, N. (1991). Man and nature relationship reflected in the Sukla Yajurved. In V.N. Jha (Ed.), *Proceedings of the National Seminar on Environmental Awareness Reflected in Sanskrit Literature* (pp. 1-6). Poona Centre of Advance Study in Sanskrit.

Lewin, K. (1951). *Field Theory in Social Science*. New York: Harper & Row.
Lorenz, K. (1970). *On Aggression*. London: Methuen.
Michelson, W.H. (1977). *Environmental Choice, Human Behavior, and Residential Satisfaction*. New York: Oxford University Press.
Mitra, S.K. (Ed.) (1972). *A Survey of Research in Psychology*. Bombay: Popular Prakashan.
Moos, R.H., & Lemke, S. (1984). Supportive residential setting for older people. In I. Altman, M.P. Lawton, & J.F. Wohlwill (Eds.), *Elderly People and the Environment: Vol 7. Human Behavior and Environment: Advances in Theory and Research* (pp. 159-190).New York: Plenum.
Moulik, T.R. (1981). Psychology of Poverty in India. In Udai Pareek (Ed.), *A survey of research in psychology*, 1971-76 (Part 11, pp. 655-681), Bombay: Popular Prakashan.
Nagar, D. (1998). *Human Reactions to Crowding*. Jaipur: Printwell Publishers.
Nagar, D. (1999). Environmental stress and human behavior. In G. Misra (Ed.) Psychological Perspectives on Stress and Health. New Delhi: Concept Publishing Company.
Nagar, D. (2005). Human rights of poor living in slums: A psycho-social analysis. In S.N. Chaudhary (Ed.), *Human Right and Poverty in India: Theoretical Issues and Empirical Evidence (Vol 5)*. New Delhi: Concept.
Nagar, D., & Mishra, M. (2001). Slum rehabilitation in Bhopal: Its psychological consequences. *Indian Cooperative Review*, 38, 171-183.
Nagar, D., & Sharma, A. (2005). Influence of community living and environmental stressors on health and well-being. In A. Dalal (Ed.), *Social Dimensions of Health and Well-being*. Jaipur: Rawat Publications.
Newman, O. (1972). *Defensible Space*. New York: Macmillan.
Osmond, H. (1957). Function as a basis of psychiatric ward design. *Mental Hospitals*, 83, 235-245.
Padaki, V., & Vyasulu, V. (1981). Towards environmental psychology. In Udai Pareek (Ed.), *A Survey of Research in Psychology*, 1971-76 (Part 11, pp. 618-654). Bombay Popular Prakashan.
Pandey, J. (2004). *Psychology in Indian Revisited: Developments in the Discipline (Vol. 1-3): Applied Social and Organizational Psychology*. New Delhi: Sage.
Pareek, U. (Ed.) (1981). *A Survey of Research in Psychology*, 1971-76 (Part 1,11), Bombay: popular Prakashan.
Paulus, P.B. (1987). *Prison Crowding: A Psychological Perspective*. New York: Springer-Verlag.
Paulus, P.B., & Nagar, D. (1987). Environmental influences on social interactiona and group development. In C. Hendrick (Eds.). *Review of Personality and Social Psychology*, 9, 68-90, Beverly Hills: Sage.
Proshansky, H.M., Ittelson, W.H., & Rivlin, L.G. (1976). *Environmental Psychology* (2nd ed.). New York: Holt, Rinehart & Winston.

Rao, T.V. (1981). Psychology of population and family planning. In Udai Pareek (Ed.), *A Survey of Research in Psychology*, (Part 11, pp. 714-746). Bombay: Popular Prakashan.

Rapoport, A. (1980). Environmental preference, habitat selection and urban housing. *Journal of Social Issues*, 36, 118-134.

Rapoport, A. (1985). Thinking about home environment: A Conceptual Framework. In I. Altman and C.M. Werner (Eds.), *Home environment* (pp. 255-286). New York: Plenum Press.

Rivers, W.H.R. (1905). Observations on the senses of the Todas. *British Journal of Psychology*, 1, 321-396.

Russell, J.A., & Ward, L.M. (1982). Environmental psychology. *Annual Review of Psychology*, 33, 651-682.

Sandhu, R.S. (1987). Not all slums are alike: A comparison of squatter housing in Delhi and Amritsar. *Environment and Behavior*, 19, 398-406.

Sinha, A.K.P. (1972). Military psychology: A trend report. In S.K. Mitra (Ed.), *A Survey of Research in Psychology*, (pp.238-265). Bombay: Popular Prakashan.

Sinha, D. (1972). Industrial psychology: A trend report. In S.K. Mitra (Ed.), *A Survey of Research in Psychology*, (pp. 175-237). Bombay: Popular Prakashan.

Sinha, M.M. (1972). Physiological and comparative psychology: A trend report. In S.K. Mitra (Ed.), *A Survey of Research in Psychology* (pp. 338-361). Bombay: Popular Prakashan.

Skinner, B.F. (1953). *Science and Human Behavior*. New York: Appleton Century Crofts.

Sommer, R. (1969). *Personal Space: The behavioral Basis of Design*. Englewood Cliffs, NJ: Prentice-Hall.

Sommer, R. (1974). *Tight Spaces: Hard Architecture and how to Humanize it*. Englewood Cliff, N. J,: Prentice-Hall.

Stokols, D. (1972). On the distinction between density and crowding: Some implications for future research. *Psychological Review*, 79, 241-254.

Stokols, D. (1978). A topology of crowding experience In A. Baum & Y. Epstein (Eds.), *Human Response to Crowding*. Hillsdale, NJ: Erlbaum.

Stokols, D., & Altman, I.(Eds.) (1987). *Handbook of Environmental Psychology*. New York: John Wiley.

Stokols, D., & Novaco, R. (1981). Transportation and well-being: An ecological perspective. In I. Altman, J. Wohlwill, & P. Everett (Eds.), *Human Behavior Environment: Advances in Theory and Research* (Vol 5). New York: Plenum.

Stokols, D., & Shumaker, S.A. (1981). People in places: A transactional view of setting. In J.H. Harvey (Ed.), *Cognition, Social Behavior and the Environment* (pp.441-488), Hillsdale, NJ: Lawrence Erlbaum.

Sundstrom, E. (1978). A test of equilibrium theory: Effects of topic intimacy and proximity of verbal and non verbal behavior in pairs of friends and strangers. *Environmental Psychology and Non Verbal Behavior*, 3, 3-16.

Tripathi, M. (1993). *Environmental Change and its Psychological Impact: A Case Study of a Village*. Master's thesis in Psychology, Gorakhpur Univesity, Gorakhpur.

Tripathi, L.B. (1997). A theoretical perspective on Man-environment relationship. In U. Jain (Ed.). *Perspectives on Environment and Behaviour.* (pp. 28-37). New Delhi: Har-Anand Publisher.

Tripathi, P.S.N. (1997). *Pariyavaran Monavighan*, Motilal Banarasidas Publications, Varanasi.

Tuan, Y.F. (1974). *Topophilia: A Study of Environmental Perception, Attitude and Values*. Englewoods Cliffs, NJ: Prentice-Hall.

Wicker, A.W. (1979). *An Introduction to Ecological Psychology*. Monterey, CA: Brooks/Cole.

Wicker, A.W. (1987). Behavioral settings reconsidered: Temporal stages, resources, Internal dynamics, context (pp. 613-653). In D. Stokols & I. Altman (Eds.), *Handbook of Environmental Psychology*, New York: John Wiley.

Wohlwill, J.F. (1973). The environment is not in the head! In W.F.E. Preiser (Ed.), *Environmental Design Research* (Vol. 2) (pp. 166-181). Stroudsburg, PA: Dowden, Hutchinson, & Ross.

Yankey, W.C. (1972). Architecture, interaction and social contact: The case of a large scale housing project. In J.G. Wohlwill & D.H. Carson (Eds.), *Environment and the Social Sciences: Perspectives and Application* (pp. 126-153). Washington, DC: American Psychological Association.

2

METHODOLOGICAL ISSUES AND THEORETICAL PERSPECTIVES IN ENVIRONMENTAL PSYCHOLOGY

One fundamental assumption shared by social scientists to gain knowledge pertains to collection of facts that reveal the principles of behaviour by using scientific methods. In other words objective investigation of empirical phenomena is one the major objective of science for obtaining true knowledge. By using appropriate scientific methodology a great deal of order in the universe is discovered. To gain knowledge about environment-behaviour relationship the scientific method is adopted. One of the major objectives of science is formulation of theory to explain facts. In this chapter a brief mention on various methodological issues will be made and application of the major methods to attain knowledge about environment-behaviour relationship will be discussed. The nature and functions of theories in environmental psychology will be discussed. Finally, major theoretical perspective developed to explain the links between environment and behaviour would be examined.

Methodological Issues in Environmental Psychology

The ultimate goal of research is development of a theory that explains the cause effect relationship found among variables. However, there are various problems and a number of methodological issues that can threaten the validity of conclusions drawn by a researcher. Furthermore, some threats to validity questions the theoretical explanation given by the researcher about causality. It is important to consider these issues in planning and evaluating research in environmental psychology. Broadly viewed, the term validity refers to the extent to which the researcher's conclusion is true or not and that it corresponds with reality. A brief summary of four different types of validity applicable to design methodologies is presented below (Cook and Campbell, 1976):

Internal Validity

Internal validity is concerned with the logic of relationship between the independent and dependent variables. In other words internal validity determines whether the findings of the research permit causal relationship among variables and that the relationship is non-spurious in nature. The problem of internal validity is high in relational and descriptive studies relative to experimental research because there is a greater possibility that other variables are confounded with the independent variables. The confounding error is one of the biggest threats to validity in research. This type of error may distort the interpretation of the results.

Construct Validity

The issue of construct validity is related to the extent to which the results of the study support the theory behind the research. Often, in such research the investigator assumes that various aspects of the particular theory have are been tested by the variables being studied.

External Validity

A third dimension on which research is evaluated is external validity. The concern of the researcher is whether the results of the study can be generalized across situations: different subjects, settings, time frame and so forth. In environmental research when the study is conducted in the experimental setting the external validity is low because the laboratory conditions are different in many aspect from the natural settings.

Statistical Validity

This type of validity is very similar to internal validity. The issue addressed in this validity relates to the question regarding the truthfulness of the cause and effect relationship between independent and dependent variables. In other words whether the relationship obtained between independent variable and dependent variable is accidental or there exists true causal relationship between independent and dependent variables. In statistics cause and effect relationship is not proved. Instead, the probability of relationship with regard to level of conclusion derived between independent and dependent variables are ascertained.

Research Methods in Environmental Psychology

The research methods frequently been used in other areas of psychology are used in environmental psychology. These methods are subsumed under two broad categories—Experimental and non-experimental research. The difference between experimental research and non-experimental research is based on the degree of control that the researcher has over the subjects and conditions under which the study is conducted. The major difference is in the ways these methods are put into use in environmental psychology relative to other branches of psychology. The two major methods frequently used to study environment behaviour relationship are different but are not mutually exclusive. Some studies may have characteristics of more than one type, although it is probable to classify a study as being primarily labelled in one category on the basis of the major objective of the proposed study.

Non-Experimental Research **versus** *Experimental Research*

To distinction between experimental and non-experimental

research is essentially based on the usage of some key words like manipulation and assignment *versus* observation (McBurney, 2002). In this section major method utilized in both the broad research categories, their advantages and limitations and their application in environmental psychology are briefly presented below.

Non-experimental Research

Some of the major methods used in this category of research are: descriptive, relational and observational research.

Descriptive Method

One of the major objectives of descriptive study is to answer to the question that relates to—"how things are". The descriptive research serves three important functions. First, it provides groundwork of those aspects of environment and behaviour with which we are unfamiliar. When encountered with a situation that is ambiguous, the first thing the research does is to orient oneself by making use of some reference points so as to reveal various aspects of the situation. The second function is to examine the boundaries of the research problem once the nature of various aspects of the environment context and its impact on human behaviour is revealed. Finally, the inferences drawn after conducting the descriptive research can be tested in relational and experimental research. In a typical descriptive study the objective is not to draw causal inferences and relationship between measures. The statistical analysis is also not performed in descriptive studies. However, environmental phenomena are studied comprehensively by utilizing descriptive methods. According to Proshansky (1972) the environmental psychologists must be concerned with searching out the dimensions and more specific properties of phenomena involving human behaviour in relation to physical settings.

The exploratory study of Barker (1968) attempts to compare the behavioural settings in two towns. In this study the researcher was not able either to manipulate any aspects of the towns being compared nor the study permitted the investigator to randomly assign towns to different conditions. However, the findings of the study do provide interesting information about various environmental and social aspects of the behavioural settings.

Relational Method

Relational research tells us "how things are in relation to other things". One basic feature of science is to view facts not as isolated or separate but as connected in some systematic and logical way. The important functions that are served by relational research are: First, is to examine the relationship between different variables. In other words the degree and strength of correlations between various variables are ascertained by utilizing the relational methodology. A second function of relational method of scientific inquiry is that it allows us to examine the long-term patterns of future trends. By calculating the correlations among variables does not make the research relational. In correlational research the researcher does not manipulate environmental variables independently. In this type of research relationships are studied among variables, however, it is difficult to ensure strong causation. In other words internal validity of correlational research is low. However, these studies are primarily used to examine the naturally occurring events.

Field Studies

In Field studies valuable information pertaining to the effects of variations in environmental conditions on dependent and outcome measures are examined. However, in these studies researcher is not able to randomly assign respondents to different conditions and is not able to manipulate the independent variables. Thus, the problems of field studies are related to low internal validity and difficulty is deriving causal inferences. The details of field studies are given in the preceding chapters on environmental stressors and crowding. To reduce the threats to internal validity, some researchers have conducted quasi-experimental studies in the field settings (Cook & Campbell, 1979; Christensen, 1977). By utilizing the quasi-experimental design in field settings the ability to draw causal inferences can be increased.

Observational Research

Observational research methods occupy most prominent role in environmental psychology. Careful observation of naturally occurring behaviour might be helpful in formulating fruitful hypothesis and better research questions for establishing causality between measures. In this method the researcher simply observes and records ongoing behaviour but does not attempt to change it or influence it. The major methods utilized in this form of research are—naturalistic observation and participant observer research. In *naturalistic observation* the research is conducted in such a way that the behaviour of the respondents are disturbed as little as possible. This type of observational research is also called *unobtrusive* which means the behaviour under investigation is not obtrude on by the researcher. Similarly, the term *non-reactive* research is also sometimes used for naturalistic observation. The presence of the researcher is not noticed by the respondents or the respondents are not aware that they are being studied and, therefore, do not react to the presence of the observer.

Experimental Research

Experimental research is defined in terms of controlled manipulation of independent variables. The understanding of the notion of control is important in experimental studies. Usually, the effect of a particular independent variable is examined and compared against some standard. In a typical experimental study an experimental introduces treatment to one group of subjects (Experimental group) while the other group (Control group) is not administered treatment. The scores on dependent measures for both the groups are compared to examine the treatment effect. However, it is essential to ensure that both the groups are essentially similar on major demographical variables before the treatment is introduced to the experimental group. One of the foremost functions served by the experimental research is that we are in a position to draw causal inferences. In other words we can demonstrate that variable X actually precedes variable Y in time.

Laboratory Experiments

In a typical laboratory study the effects of environmental variables on different dependent and outcome measures are examined in the controled conditions. The studies on crowding and noise (Baum & Paulus, 1987; Glass & Singer, 1972; Nagar, 1998; Jain, 1987) discussed in chapter five of this volume in detail are examples of experimental studies. For instance, the effect of degree of control on reaction to unpredictable noise was examined in the laboratory by Glass and Singer (1972). The subjects were provided with an opportunity to terminate the noise by using the response button. The results of this study revealed that subjects who were provided with an opportunity to control the noise were less affected by noise relative to their counterparts who were not given the freedom to terminate the noise. The study of Nagar and Pandey (1987) was conducted in the controlled laboratory setting. In this study the effects of crowding and noise on task performance and affect was examined. The degree of crowding was manipulated by increasing or decreasing the size of the group (low density *vs.* high density) in the laboratory room and the three levels of noise was manipulated by increasing or decreasing the volume of the tape recorder in which noise was recorded. The environmental condition of the laboratory room was kept constant across all the six experimental conditions. The results of this study reveal that the performance on the complex task was significantly lower in the condition of high density relative to low-density condition. In contrast, simple task performance was not affected with the variation of density and noise.

The results of the laboratory experiments have

contributed significantly in the development of environmental psychology. However, the experimental studies are criticized primarily because the setting is artificial in nature and generalization of results across other setting is weak. In other words experimental studies are low on external validity and low on mundane realism. In a typical experimental study the subjects are exposed to the treatment conditions (density, noise) for a short time interval relative to naturalistic settings. The time interval in which the subjects are exposed to environmental stressors in the laboratory experiments is very short compared to field setting and thus, the possibility of the degree of habituation or adaptation to the environmental stressors is different.

The limitations discussed above are inherent in the laboratory experiments are reduced by designing *field experiments*. In designing field experiments independent variables are manipulated and the subjects are randomly assigned to treatment conditions in naturalistic setting.

Field Experiments

The field experiments are conducted in the naturalistic setting. The methodology requires manipulation of independent variables and random assignment of respondents to different experimental conditions. A study on territoriality designed by Edney (1975) illustrates an example of field experiment wherein the effect of environmental phenomena was studied in a systematic and casual manner in a naturalistic setting. The dormitory rooms were used as laboratory by the researcher. The researcher randomly assigned half of the respondents to their own room (resident condition) and half to other students as "visitors". The results of this study revealed that respondents experienced more control in their own territory and perceive their turf as more pleasant and private relative to when visiting the territory of others. Thus, in the present experiment external validity and experimental realism were enhanced because the setting was real.

Simulation

In many conditions simulations are preferred because it is difficult to design field experiments but the researcher is interested to increase both the experimental realism and external validity. In simulation cause and effect relationship is examined between environmental variables and outcome measures. Moreover, experimental rigor is maintained and realism and external validity is enhanced in simulated studies. The study on crowding conducted by Baum and Greenberg (1975) is a good example of simulated research work. In this study the researcher used two confederates and a subject. The technique of deception was employed and a situation was created that gave an impression to potential subjects that either a small or a large number of others would be participating in the experiment. The results of this study were found to be very consistent than those obtained by other researchers. Another form of simulation used by researchers in crowding studies involves the use of model rooms. The methodology typically employed by the investigator is to construct a scale model of the room. The characteristics of the settings are then manipulated by moving partitions. The potential subjects are then asked to place people represented by clothespins in the room until they feel discomfort due to crowding in the room (Bell & Barnard, 1977; Desor, 1972). Epstein and Baum (1978) have noted that the use of model room in simulated studies are only partially successful in replicating the previous results. The subjects may lack ability to accurately experience the setting that they are judging because the setting is not only small in size but also tactile and olfactory stimulation is missing in the model room. Finally, it is also possible that subjects may not feel involved with the situation presented in the model room.

A number of studies have used photographic slides of a wide varieties of naturalistic setting and asked the respondents to rate the slides (Kaplan, 1974; Wohlwill, 1976; Paulus, Nagar & Camacho, 1991). For instance, the study of Paulus *et al.* (1991) attempt to examine the reactions of US Army families living in mobile home parks and apartment complexes to their housing. The respondents were asked to rate 13 photographs on housing complexes on a 5-point likert type scale of attractiveness. These pictures represented the full range of environmental quality. These environmental evaluations were used as indicators of environmental quality. In addition, the two senior researchers and their four assistants also gave their independent assessment quality on various dimensions of environmental quality. The interjudge reliability and construct validity of environmental quality was found to be very high. Furthermore, there was a high correspondence between researchers measure of environmental quality and picture ratings of residents. Finally, measures of environmental quality was found to be related to more positive evaluations of apartments, but had little impact in mobile homes. In other words slides offer several advantages as a simulation of the real world environment. The pictures are easy to present to a small or a large group. Furthermore, they allow a wide varieties of scene to be presented at one point of time with high predictive validity. Thus, one can conclude that slides constitute accurate representation of the human made and natural environment and can, therefore, be used with high degree of accuracy.

Measurement Techniques

In the preceding section methodological issues and methods for collecting information about behaviour in the environment are discussed. A brief overview on some of the major measurement techniques frequently used in environmental psychology is presented.

Self-Report Measures

Survey techniques are popularly used in which respondents are asked to provide information about their attitudes, behaviours and feelings in relation to various aspects of their physical and social environment. The self-report procedures include structured and unstructured questionnaires, interview schedules, rating scales and projective methods.

The self-report measures are often constructed by the researcher to meet the specific needs and are consistent with the major objectives of the study. For instance, a comprehensive questionnaire was designed to assess reactions of army families to their housing environment. Army families residing in mobile home parks and apartments were asked to rate their parks and apartment complexes along various dimensions. Some of the items on which the rating were done included attractiveness, noise, friendliness, recreational facilities, services, management, maintenance, closeness to shopping, crowding, people and safety. In a similar vein individual home were also assessed on a range of dimensions like spaciousness, privacy, adequacy for needs and so forth (Paulus, Nagar, & Camacho, 1991). Thus, based on the specific requirement of the large-scale project carried out near the army base, the self-report measures were designed which were suitable for army families (Paulus *et al.* 1991). In this particular study pilot testing was carefully done so that items that were ambiguous or were not understood by the potential respondents were either deleted in the final questionnaire or were suitably modified. Besides appropriate statistical were carried out to ensure reliability of the measuring instruments.

A number of problems are encountered when self-report measures are designed to meet the needs of a specified population or specific study. The researcher should appraise the respondents about various concepts and terminologies frequently used in the questionnaire so that every respondent may conceptualize the concept similarly (e.g. crowding, personal space). In other words if respondents derive different meaning about the technical terms or there is a mismatch between the researcher meaning of the concept and respondents differential ideas about the same concept the results obtained will be contaminated. Another common problem arises when self-report measures are used to study the needs of a specific population. For example, it is difficult to compare the results if measures used by researchers are different. This may be one reason why the results obtained by researchers differ.

A number of environmental researchers have used some

unique self-report techniques. A study of Michelson and Reed (1975) attempts to examine how people spend their time in various settings. The researcher developed time budget or activity log technique to obtain accurate data on how much time is spend by respondents in performing each of the various activities listed in a self-report format. This technique is more economical and accurate in understanding the activity pattern of respondents across some specified time frame relative to observation or interview techniques. Another form of self-report measure is called cognitive mapping. A variety of procedures are used to understand how people construct the map in their mind and the images are transported in the piece of paper. Thus, utilizing this procedure sufficient information is gained about how people code spatial information about their everyday environment. Thus, adopting the procedure of cognitive mapping a number of studies have analyzed how respondents of different demographic characteristic perceive their college campuses, local neighbourhoods and even nations.

Questionnaires

Questionnaires are widely used self-report paper and pencil technique of gathering scientific information from the subjects. Often the purpose of questionnaire is to determine how respondents feel about a particular issue such as Three Mile Island nuclear accident. For instance, some of the items were constructed to determine who evacuated the area and for what reasons. Furthermore, through questionnaire survey an opportunity is provided to examine the correlations among the respondents response and to look for possible patterns of causality.

Designing a questionnaire is complex and the researchers have to take into account many considerations. Frequently, researchers use standardized questionnaire rather than redesigning their own questionnaire. For instance, to measure the social climate Moos and Gerst (1974) constructed and thereafter standardized a University Residence Environmental Scale. The social climate scales assume that social institutions have personalities representative of the type of activities and interactions that occur in the setting. The scale comprises 90 to 100 items and incorporates seven major dimensions. A modified version of this scale entitled indwelling crowding experience scale comprising of seven dimensions was used in residential setting in the Indian context (Nagar, 1998).

The standardized questionnaires constructed by environmental psychologists are very few in number. Thus, researchers have to rely on non-standardized questionnaires. In most cases the researchers scan the available literature and depending on the context and research objective of their study construct appropriate items and elicit information from the respondents. In many cases exploratory approach is adopted and appropriate open-ended items are constructed to gather range of useful information. These strategies suffer from various limitations. For instance, it is possible that the respondents may not wish to give detailed information about the content domain or it is time consuming for the respondents to provide comprehensive details about the phenomena under investigation. Moreover, expressive power of the respondents may limit the range of dimensions on which they are asked to respond. In other words open-ended items require more effort from the respondents and are more difficult for less articulate respondents to answer. Furthermore, it is often difficult to categorize information provided by the researcher appropriately, especially in cases when the range of answers and narrations given by respondents are wide. Although, content analysis computerized procedures are available in organizing the responses given by subjects on open-ended questions (McBurney, 2002). To deal with some of the above mentioned problems inherent in open-ended questions, some investigator use open-ended items in pilot studies and thereafter based on the results obtained they construct closed ended, multiple choice or other appropriate type of questions for categorizing the responses.

In short, the flexibility of open-ended items makes them more useful for exploratory and small-scale studies. In contrast, closed ended questions and standardized instruments are more beneficial for large-scale research projects. Some researchers incorporate both open-ended and closed ended items in their research project to gain advantage of both types of questions.

Interview Techniques

Interview techniques may comprise of structured or unstructured items. According to Campbell (1950) the unstructured items offer the respondents the opportunity to elaborate their answers to express their behaviour and feelings spontaneously. In contrast, structured items have clear-cut response options. The researcher may use tape recorder to record verbatim response from the respondents when interview technique is unstructured in nature. The interviewer may intervene if needed to seek clarification or to give specific directions. The famous study of Lynch (1960) on images of city is an example of the utility of unstructured interview technique. In contrast, the items of structured interview are prepared keeping in mind the specific research objectives. The case study conducted by Cooper (1971) on low-income housing projects utilizes a structured interview technique. In this research project comprehensive questionnaire comprising of various background information about the respondent and specific questions about the needs and values of the respondents were obtained.

Rating Scales

One of the most popular methods of measuring and recording various aspects of behaviour are rating scales. In environmental research some of the most frequently used rating scales to assess the feelings are Nowlis Mood Adjective Checklist, subjective stress scale constructed by Emiley (1975), preferred level of environmental stimulation designed by Zukerman (1971) and description of architectural form (Kasmer, 1976). The details of these scales are provided in the book authored by (Fischer, Bell and Baum, 1984).

The 5-point equal appearing Likert type scale rating scale are frequently used to assess various aspects of environment and behaviour relationship. Another commonly used bipolar adjectives scale can be designed for obtaining respondents' reactions in response to various aspects of the environment. The subjects are given a list of adjectives depicting one extreme of an attribute to the other. Furthermore, some of the semantic differential rating scales (e.g. Osgood, Suci & Tannenbaum, 1957) are also used to examine how well the respondents describe the setting. Three conceptual components are represented in the scale that comprised of a large number of bipolar adjectives. These were related to evaluative aspects (good—bad), potency (e.g. strong—weak) and activity (e.g. active—passive).

Ways of Knowing about Environment-Behaviour Relationships

Existing methods to learn about environment-behaviour relationships can be broadly divided into two categories—empirical and non-empirical methods. Empirical methods used to obtain information about environment-behaviour links are primarily based on experience. However, philosophers of science have argued that the method of science is actually one of the methods to gain knowledge (Reilly, 1970).

Non-empirical Methods

The most frequently used methods of gaining knowledge are method of tenacity, method of authority, a priori method, mystical approach and method of religion (Rosenthal & Rosnow, 1984). For instance, the *method of tenacity* advocates that familiarity is enhanced if we hear some statements over and over again. We start believing in such statements because it brings peace to our mind. Similarly, in *religion* the basis for reducing uncertainty is faith, tradition, revelation and experience. In the mystical mode, knowledge is gained from prophets, divines, gods, mediums and other supernaturally knowledgeable scholars. To large extent ritualistic and ceremonial procedures to the target person is applied.

Empirical Methods

The major focus of research conducted in the area of environmental psychology accepts the notion of determinism.

In other words the scientists believe the causal linkages between events and those scientific methods can help us to predict one event from the other. For instance, using the scientific methods the environmental psychologists can predict that with the increase in the levels of illumination and noise (cause) the productivity (effect) of the workers will reduce. That entire phenomenon (e.g. supernatural events) which are not observable and whose nature is indeterminate or is unpredictable fall outside the perview of scientific inquiry.

The scientific methodology utilized in the physical science is followed in research studies in environmental psychology. The research is based on explicit rules and procedures against which claims of knowledge are evaluated. Logical validity and empirical verification are two essential criteria employed by scientist to evaluate claims of knowledge. The scientific process comprises of seven major stages: problem, hypothesis, research design, measurement, data collection, data computation and generalization. The research process is cyclic in nature with the starting point being selection of problem and the concluding point being a tentative empirical generalization. The cyclic process continues indefinitely and ending of one cycle after generalization the new cyclic process continues which reflects the progress of scientific discipline. The empirical and scientific method starts with the selection of problem where the record of observations is possible. The researcher formulates his hypothesis about the expected relationship between independent and dependent variables. The hypothesis is subjected to verification. The observable data are to be made public so that other researchers can replicate the same study by adopting the same methodology and tools and verify the results. In case the hypothesis is rejected the researcher must either modify his previous hypothesis or may test new hypothesis and verify the utility of the hypothesis. After a large number of empirical and observable facts are gathered and the relationship between facts are known it is desirable that the research may proceed to a more theoretical and abstract level so as to explain the observed relationships.

Nature and Function of Theories in E.ivironmental Psychology

Before the theoretical perspectives are discussed it is essential to throw some light on the fundamental difference between laws, theories and model. Laws are statements that express simple observable relationships between phenomena. Relative to law a model is more complex and abstract. While the theory tends to involve more abstract concepts and while the constructs are broader in scope. Finally, the model is an intermediate link between easily observable empirical laws and the formulation of theory.

A theory state suspected relations between concepts. Theories guide research activities. Theories are tentative and temporary ways of observing the phenomena that await further test. A good theory is formulated when different researchers divide the problem of investigation into smaller and manageable research questions and subsequently seek solution by integrating and providing explanation for the results obtained. Thus, theory serves to organize research. Separate researchers working on a similar pı oblem may coordinate their research efforts within a framework of a shared theory. There are three major functions of sound theories. First, theories are helpful in predicting the relationship between variables. For example, if we know that certain conditions of home crowding lead to aggression, we can reduce aggression by changing the crowded conditions. If the privacy reduction theory states that providing desired quota of privacy to family members mediates the relationship between crowding and aggression. To reduce aggression in the home provisions of privacy to family members is helpful. A second function of a good theory is to summarize large amount of data. If we have a good theory of stress we can summarize a large body of studies dealing with adverse effects in few theoretical statements. Moreover, prediction of events is also possible even when empirical data on stress are not readily available. Finally, generalization of results across different conditions is possible within a well-grounded theoretical perspective. For example if stress theory states that high level of stress increases aggression, then we can generalize the notion that any condition that produces stressful conditions like crowding, noise, temperature and so forth may also produce aggression. In case the empirical findings fails to suggest that a theory generalizes well then the theory should either be modified or rejected in favour of another theory which is more parsimonious. Apart from the three major functions theories are helpful in stimulating new research ideas that can be subjected to empirical investigations. Furthermore, the theorist may invent new concepts and predict new links between concepts. Finally, good theories are highly coherent, logical and are high on internal consistency. They are able to integrate existing thoughts and empirical data taking into account both agreement and disagreement among observers. As been practiced in all fields of science theories of environmental psychology are constantly subjected to evaluation and verification. Of the two theories one with relatively high degree of predictive validity, better generalizability and better capability of summarizing large data set is preferred.

As compared to other fields of psychology the discipline dealing with environment and behaviour is relatively new. The field of environmental psychology not only lacks focused direction but is also deprived of sufficient data set on which the sound foundation of good theories is based. Within the backdrop of the deficiencies mentioned above elaborate theories that meets the criterion of prediction, summarization, generalization and additional research stimulation are not developed. At this point of time scholars in the field have attempted to develop restricted theories whose range of applicability is limited to its predictive domain. Attempts are been made by researchers to develop models depicting the environment-behaviour relationship. Finally, only limited attempts are made to construct theories that are comprehensive and contextual in nature. In this chapter attempt has been made to discuss some dominant theories that are frequently been used and whose predictive powers are limited to specific situations. In other words, some theories developed across the period of time which focus on environment-behaviour relationship will be discussed. These are: (1) Arousal theory;

(2) Stimulus overload theory; (3) Adaptation approach; (4) Behavioural constraint theory; (5) Ecological framework and (6) The stress approach. The ecological perspective is more useful when group behaviour is considered while other approaches are more useful when the level of analysis is individual.

The Arousal Approach

The term "arousal" or "activation" refers to a person's general state of alertness or excitation, both physiological and psychological (Scott, 1966). The physiological components of high arousal which can be measured precisely includes heightened autonomic activity which is indicated by increase in blood pressure, blood flow in peripheral vessels, heart rate, respiration rate, skin temperature, muscular tension and adrenal secretion. The levels of arousal can also be measured at the behaviour level by monitoring the motor activities or psychologically by the help of self-reported inventories. From the neuro-physiological perspective reticular formation centre of the brain is activated by heightened brain activities (Hebb, 1972). Similarly, Duffy (1962) has conceptualized arousal in terms of certain type of brain activities. The level of arousal ranges from drowsiness through alertness to excitement or heightened wakeful activities on a continuum anchored at one end from drowsiness and to heightened wakeful activities on the other. In many studies arousal is treated as an intervening or mediator variable to explain the relationship between environment and behaviour.

The principal hypothesis that relates arousal to performance holds that people perform optimally when the level of arousal is of moderate level Hebb (1949). The Yerkes-Dodson law states that the performance is maximum at intermediate level of arousal. The performance progressively deteriorates when either the level of arousal is either low or high. Furthermore, the "inverted-U hypothesis," shows that performance declines in either direction as level of arousal departs from the optimal range. While minor deviation from the moderate degree of arousal towards either relaxation or excitement tends to produce marked decrements in performance. Moreover, the "inverted U type" of relationship between arousal and performance depends on the complexity level of the tasks. The optimal level of arousal is lower for the complex tasks than for the tasks that are simple in nature (Yerkes & Dodson, 1908). According to the arousal hypothesis, the impact of physical environment on task performance depends upon the demands of the task. The performance is improved if the demand imposed by the task creates an arousal into the optimal range and *vice versa.*

The importance of arousal from the environment-behaviour perspective is noted in many studies which tend to deal with environmental stressors like crowding, noise, air pollution (Nagar & Pandey, 1987; Kahneman, 1973; Hebb, 1972). A number of studies have noted that conditions creating heightened arousal resulting from stressful conditions like crowding and noise are explained with regard to observed behavioural effects (Broadbent, 1971; Seta *et al.* 1976; Evans, 1978). The stimulating properties of a physical environment reflect both physiological and psychological processes. For instance an intense physical stimulus, such as loud noise, bright light, or strong odour can lead to arousal for psychological reasons, including perceived novelty or threat (Sundstrom, 1986).

The arousal theory gets support largely from the studies conducted in the controlled laboratory setting (Nagar, 1998; Duffy, 1962). However, the theory is rarely been tested in which performance is measured under conditions of low, moderate and high arousal. There are measurement problems largely because physiological measures are not always correlated with other physiological or psychological measures. Furthermore, it is very difficult to measure arousal. One measure may indicate high arousal in a given situation other measure may show low arousal. Thus, which measure to choose in predicting behaviour becomes problem. Besides these shortcomings the arousal framework is fairly accepted as an explanation for the impact of range of physical stressors like crowding and noise on performance.

Behavioural Constraint Theory

The behavioural constraint perspective is derived largely from Brehm's theory of psychological reactance and Sommer's conceptualization of personal space. The term constraint means that some type of interference is causing goal blockage. The interference leading to goal blockage may be either in a form of belief or may actually cause impairment of task performance resulting from the constraint posed by the environment. Once the individual perceives that the environmental events are interfering with goal attainment individual feel uncomfortable or experiences negative feelings. He attempts to restore or regain his freedom by removing the obstacles. This phenomenon is known as psychological reactance. Brehm (1966) states that maintenance of freedom of choice is an important motivating factor in human behaviour and perception. He argued that people are disposed to maintain or restore freedom when it is threatened and that individual's reaction to a setting is dependent on his success at accomplishing this goal. A number of studies have reported that even anticipating that some environmental factors may restrict our freedom a psychological state of reactance is set in and adverse social and psychological consequences are seen (Worchel & Teddlie, 1976). Sommer's (1969) notion of personal space is related to the construct of reactance. Personal space is conceived as an invisible "emotionally charged" zone surrounding each person, "into which intruders may not come". Personal space presupposes the existence of individual's needs and situational norms which establishes the bound of comfortable interactional distance. Violation of spatial needs and norms create within the individual a reactance like pattern of emotional distress and the person adopt certain ways of adjustment in order to preserve his personal space. This theory of reactance has received support from other scholars (Stokols, 1972; Proshansky *et al.*, 1970) while explaining the crowding phenomena. An important assumption underlying behavioural constraint analysis of crowding is that the mere anticipation of social interference from others in one's immediate area plays an important role in determining the psychological and behavioural variables as well as spatial dimensions of an area. The findings of many studies are in consonance with the above mentioned assumptions.

Three basic steps are conceived while explaining the applicability of the behavioural constraint model. These are related to perceived loss of control, state of reactance and feeling of helplessness. For instance, in response to excessive or undesirable stimulation attempts made by the individual to regain the control over the environment are unsuccessful a psychological state of helplessness becomes evident (Seligman, 1975). If repeated attempts to restore control over the environmental event results in failure, individuals tend to withdraw or abandon their efforts and give up trying and perceive they are helpless and negative affective states like depression and negative consequences on behaviour are generated. A number of studies conducted in the area of environment and behaviour reveal that conditions where subjects are provided actual or even perceived control over the situation negative affects are reduced to a large extent. For instance, Glass and Singer (1972) instructed their subjects that they could reduce or even eliminate the level of noise if they want to by pressing the button. The subjects while performing the task did not use the button to reduce the noise. It seems that by just perceiving that they can control the aversive stimulus the adaptive cost resulting from noise was reduced. Other studies have reported evidence that perceived control over the noise and crowding have also reduced negative effects of aggression, increased helping behaviour and reduced unpleasant, undesirable effects (Donnerstein & Wilson, 1976; Sherrod, 1974; Langer & Saegert, 1977; Rodin, 1976).

With regard to the notion of control Averill (1973) discusses three categories of control. The behavioural control provides subjects an opportunity to change the disturbing environment events like pressing the button to switch off the noise. In cognitive control the participants appraises the environment as less threatening (e.g. believing that working in extreme temperature is not harmful). Finally, the decisional control provides subjects an opportunity to choose the best possible alternative. Sherrod (1974) in their study reported relative to the subjects who were given an opportunity to control either onset or termination of noise the subjects who were provided an opportunity to control both the onset and termination of noise exhibited better adaptation. Thus, it seems that more control over the situation is perceived as beneficial with regard to adaptation. The behavioural constraint model is able to explain and predict behaviour to a large extent in conditions when the component of control is used. However, if inferences are not based on control other mediators like arousal, overload and stress are needed to better explain the relationship between environment and behaviour.

Stimulus Overload Theory

Early theorists regarded components of environment as a source of distraction that was responsible for the diversion of attention from the task leading to deterioration of performance (Morgan, 1916). The assumption is that the individual's span of attention is limited and depending on the priority the task or components of a task are handled by the individual. This distraction hampers the task performance is noted by many scholars. Stimulus overload analysis has largely been based on earlier sociological theory of urban life developed by Wirth (1938). According to Wirth (1938) the urban areas can be characterized in terms of size, density and heterogeneity. These characteristics often result in the exposure of city residents to excessive levels of social and physical stimulation. The surfeit of stimulation, resulting from congestion and noise is linked to a number of consequences at the community level like a cultural lifestyle of impatience, aloofness, social isolation and elevated levels of physiological and psychological discomfort.

An extension of the earlier sociological theory was attributed to Milgram (1970). He defined overload as a situation in which the amount and rate of environmental inputs impinging on the individual exceeds the capacity to cope with them. He believes that individuals must enact certain insulative strategies under the condition of overload, if he is to operate effectively in the urban environment. The strategies of adaptation to overload are disregard to low priority input, allocation of less time to each input and the development of aloof orientation towards strangers. Thus, the individual spends less time with each social contact ad develops more superficial interpersonal relationships with others. A more detailed explanation of the environment load approach can be seen in the work of Cohen (1978). According to Cohen (1978) individuals has a limited information processing capacity and can invest limited effort to deal with environmental inputs at a given time. Situations when environmental inputs exceed the attention processing capacity, some inputs that are not or less relevant to the task at hand are ignored. When a stimulus occurs that require some sort of adaptive response, a monitoring process evaluates the significance of the stimulus and the individual decides which coping strategies is put into use. If the stimulus is intense, unpredictable or uncontrolable the adaptive capacity required is more and thus the individual is required to pay more attention. Similarly, if the inputs are uncertain more adaptive response and attention processing capacity is required. Finally, the overload model assumes that prolonged demand results in temporarily depletion of the attention processing capacity. According to the overload model, one's capacity for attention has been depleted in response to prolonged demands even subsequent small demands becomes overloading. Finally, the model states that individual's attentional capacity is not fixed but shrinks when there are prolonged demands. This shrinkage presumably increases with both the attentional overload of an activity and the duration of the activity. Evidence exists that once exposure to unpleasant or excessive stimulation has ceased, behavioural after-effects, such as decreased tolerance for frustration, errors in cognitive functioning and reduction in helping behaviour are evident (Milgram, 1970).

According to Miller (1964) a person may use several strategies when confronted with various sources of overload. These are as follows:

Filtering	: Selection of only certain signals according to priority.
Approximation	: Simplified response.
Queuing	: Letting work accumulate during period of demand and catching up during lulls.
Omission	: Failing to respond to certain signals. Error in processing. Escape from the situation.

Miller (1964) conducted an experimental study to test the applicability of the overload theory. He used a machine that presented a sequence of pictures of arrows in one of the eight positions corresponding to the point in the compass. The task was to push a button in one of the several sets of buttons that corresponded with different positions on the screen by selecting the proper button within the set. During the study the rate of display was increased to overload the subjects. The results of this study indicated that of the several strategies available, the participants utilized the filtering approach by focusing on the selected location on the screen. When the rate of input exceeded their capacity to watch even one location, they resorted to omission and ignored signals they could not process.

Application of overload model to the issues of crowding is reflected in the work of many psychologists (Desor, 1972; Zlutnick & Altman, 1972; Nagar & Pandey,1987). Altman (1975) has characterized crowding, as a condition in which interpersonal control boundary mechanisms break-down, such that achieved privacy is less than the level of privacy desired. A number of scholars have examined the overload interpretation of crowding. For instance, Desor (1972) utilized a role playing technique and asked his subjects to place miniature figurines in a model room until the room was just short of being crowded. The number of entrance to the room and presence and absence of partitions were varied. She found that more figurines were placed in the rooms where partitions were present and there were fewer external entrances. In a similar vein the study of Altman (1975) reveal that partitions and entrances are helpful in regulating social contact. Valins and Baum (1973) found that students tended to rate corridor-style dormitories more crowded than suit-style dormitories. They suggested that the former design is less efficient in shielding residents from unwanted social interaction.

A number of studies have reported that high density situations are informationally complex and are found to hinder environmental perceptions. Saegert (1973) reported that customers in the Manhattan departmental store were less able to recall the details concerning the merchandise and layout of the shoe department under high density conditions relative to low density settings. Thus, one can infer that perceptual and behavioural deficits will occur in high density situations where individuals were unable to avoid excessive environmental stimulation. Similarly, Milgram (1970) suggested disruption of social life is evident in large urban areas besides a number of perceptual cues are missed. Even in crisis the bystanders fail to help or are unconcerned with social situation leading to distress of high magnitude (Latane & Darley, 1970).

In the nutshell, overload model has proven to be helpful in explaining a variety of relationships between environment and behaviour particularly in situations where individual is overwhelmed with inputs and consequently his attentional capacity is depleted. This model has generated the possibility of lot of research in the area of environmental stress.

Adaptation Theory

A number of studies have pointed out that both excessive stimulation and under stimulation have deleterious impact on performance, behaviour and affective states. The adaptation framework discussed by Wohlwill (1974) points out that the optimal stimulation or intermediate level of arousal is ideal and is beneficial for task performance and generating positive emotional states. For example, humans dislike being with too many people in various situations. Similarly, social isolation for a long period of time makes people uncomfortable.

Wohlwill (1974) believes that optimal level of stimulation is most desirable for all types of stimulation including temperature, noise and crowding. The three categories of stimulation discussed by Wohlwill include sensory, social and movement where optimal level is considered desirable. These three categories of stimulation vary along the dimensions of intensity, diversity and patterning. For instance, conditions in which people are concentrating on some important task intense noise of tape recorder played by the neighbours seem irritating. In the same vein, people get disturbed if no sound is permitted in the room for a long period of time. Thus, optimal level of stimulation is considered best in many social situations. This optimal level of stimulation varies across persons and situations as a function of experience and length of time interval in a given environment. For instance, the threshold of tolerance for people living in metropolitan cities is high relative to people living in small villages. Similarly, people residing in high altitudes are comfortable with little oxygen as compared to people living in plains. This shift in the optimal level of stimulation is referred to as adaptation. Wohlwill (1974) defines adaptation as "a quantitative shift in the distribution of judgemental or affective responses along a stimulus continuum as a function of continued exposure to a stimulus."

Adaptation refers to processes that allow people to adjust to their environments, including changes in the perception of the environment and actions taken to cope with problems it create (Dubos, 1980). With regard to differential environmental component present in the work places evidence of perceptual and behavioural adaptation is seen. When the person is exposed to an environment for a long period of time a shift in perception toward evaluating it as less extreme is reported. For instance perceptual adaptation is seen to take place when an individual is exposed to loud noise, bright light or hot conditions. Loud noise causes a tightening of certain muscles in the ear, which attenuates the sensation of sound. Thus, involvement of automatic physiological process is evident in perceptual adaptation. In addition, the theory of adaptation reveals evidence of involvement of psychological process

(Helson, 1964; Wohlwill, 1974). An adaptation level constitutes a psychological benchmark for evaluating new conditions. For instance, a person working in noisy factory environment for a long period of time may judge an office environment as relatively quiet.

Psychologists have utilized the term defensive reappraisal, which can occur as a way of coping with stress. In this form of perceptual adaptation the environmental conditions once seen as very threatening are reevaluated as less threatening. For instance, a factory worker who in the initial stages of work fears the ill effects of air pollution may reevaluate the same situation as less harmful. This process of defensive re-evaluation helps the person to deal with the stress emanating from the consequences of air pollution and blinds the worker to its potential danger. Behavioural adaptation is seen when person working in high noise conditions putforth extra effort to concentrate on the task so as to deal with low performance. However, when the person modifies some aspects of the environment (e.g. using ear protectors to deal with noise) and is not involved in changing the responses to deal with adverse consequences the term adjustment is used instead of adaptation.

The theory of adaptation is useful because it incorporates some of the best features of the overload, stress and stimulation theories to explain the relationship between environment and behaviour. However, the theory is not able to explain the variation in adaptation level across individuals and thus generalization and precise prediction with respect to the impact of environment and behaviour becomes difficult.

Ecological Theory

The ecological perspective was first articulated by biologists in the late nineteenth century, and subsequently by sociologists. Barker (1960) applied the ecological theory to understand the human social behaviour. The ecological approach considers the environment-behaviour relationship as bi-directional and interdependent entities. The primary focus of Barker and his colleagues (1968, 79) ecological theory is behavioural setting, which exists within a defined physical structure. Before an elaborate discussion of ecological theory is made it seems essential to understand the concept, nature and major functions of behavioural setting. Furthermore, the undermanning and overmanning theories, which are derived from the behavioural setting theory, will be discussed.

Behavioral settings are time and place-bounded activities such as classroom, factory, temple. Each of these different types of behaviour settings are characterized by a pattern of behaviour that remains constant across the period of time irrespective of the change in the composition of setting members. A wide range of behaviours can occur inside a structure, which is comprised of four walls, ceiling and floor. If the cultural purpose of the setting signifies a classroom than the task carried out in this setting will be related to learning specific skills and is different from the purpose of a temple or recreational room. This cultural purpose exists because the behavioural setting consists of interdependency between standing patterns of behaviour and physical provisions. Standing pattern of behaviour represents the collective behaviour of users rather than individual behaviours. These behaviours are unique to the setting. For instance, if the behavioural setting is a classroom, then the standing patterns of behaviours would range from lecturing, taking notes, asking questions, exchanging information and so forth. Once we know about the characteristics of the behavioural setting prediction of behaviour is possible. However, a change in either the standing behavioural pattern or the physical milieu results in changes in the behavioural setting.

Barker's theory of behavioural setting is molar and contextual in nature and it emphasizes on the interdependent relationship between person and environment. These aspects of the theory are found to be of great interest to scholars working in the area of community and organizational behaviour (Oxley & Barrera, 1984; Greenberg, 1979; Rapoport, 1980; Wicker, 1979). The ecological theory of behavioural setting can be applied in assessing the layout and design of residential areas. The location of the play area, lobby, pathways, elevators, parking spaces and so forth are important design features should be given consideration to avoid congestion and ensure safety.

Each behavioural setting (e.g. classroom, coffee shop) is characterized by a pattern of expected behaviour from the occupants. The stability of these behavioural patterns derives from the setting occupant's desire to pursue certain opportunities the setting affords and consequently, the occupants' willingness to take desirable actions to endure effective functioning or dealing with threatening conditions (Barker, 1979). One pertinent question is what happens if a behavioural setting has too many or too few occupants? From the ecological psychology perspective the theory of manning (Wicker, 1979; Barker, 1960; Barker & Gump, 1964; Wicker & Kirmeyer, 1976) which is derived from the behavioural setting theory attempts to answer the above mentioned and related questions. Before the detail of the theory of manning is discussed it seems important to define the frequently used concepts around which the theory of undermanning and overmanning revolves. The term *maintenance minimum* signifies the minimum number of occupants needed to maintain the setting programmes while the maximum number of occupants, which a setting can hold, is called *capacity*. The *applicants* are prospective occupants who are eligible for membership and are willing to join the setting. *Performers* in a setting are those occupants who carry out the primary task such as teachers in a classroom. Finally, those involved in secondary roles like students in the classroom are defined as *non-performers*. Maintenance minimum, capacity and applicants are different entities for performers and non-performers in a behavioural setting. For instance, maintenance minimum for performers in the school would be the smallest number of staffs (e.g. teachers, secretaries, and dean) required for carrying out the programme. Similarly, maintenance minimum for the non-performers would be the smallest number of pupils required to keep the going on of classroom activities. The capacity for the performers in a typical classroom may be determined by social (e.g. number of teachers) and physical factors (e.g. size of the room) while for non-performers the room size is the primary determinant of capacity.

If the number of applicants in a setting falls below the maintenance minimum, then some or all the applicants are required to take more roles if the behaviour setting is to be maintained. This condition is termed as undermanning. In a similar vein, if the number of applicants exceeds the capacity, the setting is overmanned. Finally, if the number of applicants is between maintenance minimum and capacity the setting is adequately manned. If fewer persons are available than the number prescribed by the programme, then those who are present are assigned additional tasks so that setting continues to operate without interruption. The persons are expected to assume more difficult, important and varied roles. In contrast, if a greater number of people are permitted access to the setting than it can afford, some of these may be assigned redundant roles or are only given part of the role assigned to others hence, their setting claim is expected to be diluted. Wicker and Kirmayer (1976) have pointed out the consequences of undermanning and mechanism for regulating the population of the behavioural setting. Occupants in the undermanned settings are required to perform several tasks and roles. As a result their performance is expected to go down. Furthermore, admission standards are usually kept low and less attention is paid to the differences between individuals. In contrast, when the setting is adequately manned individual differences with respect to performing the various roles are noticed with respect to their ability and hence the performance standards are usually high. As compared to the adequately manned settings, in the undermanned setting each individual is valued and is given more responsibilities so that they can contribute more meaningfully to sustain the viability of the setting. While the undermanned settings have more opportunities of both success and failure hence the feeling of insecurity is high relative of the undermanned settings. When the behavioural settings are overmanned certain adaptive mechanism becomes functional so as to keep the vitality of the setting. For instance, moving into a relatively larger physical space or enlarging the physical milieu is one solution to deal with the adverse consequences of overmanning. Another regulatory mechanism includes specifying the time limit of occupant's stay in the setting and delineating stricter requirement to control the entrance process.

In general prediction from the theory of manning has been supported by the research findings. For instance, a number of studies report evidence that the size of a given type of behavioural setting is inversely related to the claim experienced in them (Barker & Gump, 1964; Wicker & Kauma, 1974; Baird, 1969). These studies were conducted in the high schools and churches and employed various measures of setting claim. Most often setting claim was operationalized as behavioural indices of participation in the setting, such as attendance, performance of operative *versus* spectator roles, hours spend in the setting and so forth. To a lesser extent, perceptions like felt obligation to attend activities and sense of belonging to the setting are also used to understand the setting claim. The results obtained from various studies conducted in small *versus* large high schools (Barker & Gump, 1964), colleges of varied size (Baird, 1969) and large *versus* small churches (Wicker & Kauma, 1974) are consistent with the manning theory and are found to be very meaningful and relevant in assessing involvement, satisfaction.

In sum the ecological theory and its derivatives are having both advantages and limitations. The theory is useful because it has been tested in real life behavioural settings. However, precise causal relationship between various predictors and outcomes is difficult, as the number of extraneous variables is not controlled. The theory is very broad in scope and hence specific prediction becomes difficult. The theory has proven to be very useful in understanding group behaviour and has generated many research ideas and questions.

The Environmental Stress Approach

The concept of stress is widely used in environmental psychology. Many aspects of the environment like crowding, noise, heat, natural disaster are stress provoking and aversive that is related to a wide range of negative outcomes which includes emotional, behavioural and physiological components. The physiological perspective of stress proposed by Seyle (1956) is often labelled as systemic stress. The behavioural and emotional components are often labelled as psychological stress. Within the perspective of environmental psychology both systemic and psychological components are combined into one theory (Baum, Singer & Baum, 1981; Lazarus & Cohen, 1977; Nagar, 1999). It is observed that both physiological and psychological stress reactions are interrelated and so they do not occur alone. The term 'stress' used in this volume will refer to the entire stimulus-response situation and 'stressor' will be used in reference to environmental component alone.

Seyle (1956) developed the physiological perspective of stress phenomenon. The general adaptive syndrome (GAS) on which his physiological perspective is based include three stages: alarm reaction, the stage of resistance and exhaustion. During the alarm reaction phase the autonomic processes are active. The body secretes adrenaline and neoadrenaline into the blood stream, which stimulates the adrenal cortex to produce various chemical substances including the secretion of anti-inflammatory corticostreoids. When the adaptive reserves are made ready, the body enters the stage of resistance to combat stress posed by the invading pathogen by setting up a feedback loop and stimulating the adrenal medulla to release more catecholamines. As the pathogen is invaded the adrenal medulla returns to normal level and adaptation is achieved. When the threats are repeated frequently at short intervals, the organism remains in a state of arousal until adaptation is attained or the adaptive reserves are depleted bringing the organism into a stage of exhaustion. At this stage psychological disfunction's occurs and the body suffers breakdown or disease.

The physiological model of Seyle (1956) has not distinguished between responses to physiological and psychological stressors. The model has also not analyzed the mechanism that tends to trigger the GAS. Besides simple autonomic stimulus-response relationship important cognitive components are incorporated in the concept of stress. For instance, not all stressful stimuli are aversive enough to evoke automatic alarm and resistance responses. The psychological perspective advocates that the stress process will begin only if the cognitive appraisal of stimuli is perceived as threatening. Evidence exists that the same stimulus may be threatening in one situation and not threatening in the other situation. It is frequently observed that individuals residing in areas where the level of heat is high will have little complains about the temperature but the same amount of heat is sufficient to generate uneasiness to individuals who live in areas where the climatic conditions are cold. In addition, cognitive appraisal of an aversive environmental event which is not physically present and is only anticipated as sufficient to elicit stressful reactions is noted by many scholars (Worchel & Teddlie, 1976; Baum & Greenberg, 1975). According to Lazarus (1966) the cognitive appraisal is a function of individual psychological factors (e.g. past experience, motivation) and cognitive aspects of the situation (e.g. degree of control and predictability of the situation).

A distinction is made between primary and secondary appraisal. Primary appraisal usually involves assessment of threat—anticipated harm, harm/loss that has already occurred or threat that can be dealt with. According to Evans and Cohen (1987) the primary appraisal of stressors depends on personal and situational factors. Personal factors influencing primary appraisal include general beliefs about self-efficacy, centrality of goals threatened by the stressors and dispositional factors. The situational factors that may influence primary appraisal include the duration of stressor, controllability of the stressor etc. If the individual makes an appraisal of threat or harm, then the process of secondary appraisal gets started. Individual evaluates his coping strategies and resources to deal with the stressor. The problem focused coping strategies involves changes in the situation directed towards reducing the adverse impact whereas; the emotional-focused coping strategies tend to alter the individual responses to the negative situation. Both these coping strategies can assume direct action like information seeking, flight or palliative activities (e.g. psychological defense mechanism like denial, drugs). Evidence exists that in stressful conditions (nuclear disaster) when direct action is not available palliative activities are more likely or probable (Baum *et al.*, 1981). If the coping attempts are unsuccessful in dealing with the adverse effects of stressor the individual enters the stage of exhaustion. However, in many situations when the aversive stimulus is presented in quick succession the stress reaction to it becomes weaker and the individual is able to adapt to the conditions. For instance, when first exposed to air pollution in metropolitan cities the individual may feel uncomfortable or may suffer from physiological symptoms and may report health related complains. However, after staying in the same city for a long period of time the individual adapts to the situation and the stimulus is no longer perceived as threatening.

To understand the impact of environmental stressors the models of adaptation and coping are found to be very useful. Adaptation to stress is both beneficial and costly. Individuals are able to withstand substantial amount of environmental demands largely because they have a broad and flexible repertoire of coping resources that allows them to maintain equilibrium or near equilibrium in the face of broad array of environmental conditions (Evans & Cohen, 1987). Stress is beneficial if stressful situations are handled successfully is noted in many studies. For instance, in many studies of crowding and noise stressful conditions are created that elevates the arousal level of subjects that consequently enhances the performance of the individual (Baum & Paulus, 1987). Even when the stressful situations are terminated the aftereffects is reported in many studies (Evans, 1987; Cohen, 1980). However, in conditions when the individual is not able to cope with the stressful conditions effectively negative consequences ranging from performance deficits, psychosomatic complaints and sometimes even psychological and physical breakdown is seen.

The theoretical model of environmental stress is found

to be very comprehensive. It has a broad range of predictive domain and the model can be effectively put into use while explaining large data set that deals particularly with the adverse consequences of wide varieties of environmental conditions like crowding, noise, heat, pollution and so forth. The stress approach has great deal of generalizability across different stressful conditions. The effects of multiple environmental stressors and their individual and combined impact on human behaviour are explained comprehensively. The environmental stress approach has generated substantial research and has raised many questions of theoretical and practical importance.

Thus, far some common and frequently used theoretical perspectives in the understanding of environment and behaviour are discussed. These theoretical perspectives are not mutually exclusive and the effects of stressors can be explained within the framework of different theories. Different mediators are selected to explain the casual connections between stress and its consequences. However, each theory selects one or two mediators to explain the results. Nevertheless, it does not mean that other potential mediators are not operative for the same data set. It is very plausible that crowding produces cognitive overload, interference, stress, and arousal at the same time in the same individual. Thus, regardless of which mediators are used a number of coping responses are likely to operate in response to stressful conditions. However, it is possible that the predictability of one mediator is better than the other in a given situation.

Fischer, Bell and Baum (1984) present an eclectic scheme depicting the environment-behaviour relationship. The model illustrate that objective environmental conditions like density, noise, exists independently of the individual. The individual may take action to change these objective conditions. The model discusses the factors related to individual differences (levels of adaptation, perceived control and personality characteristics) as well as social factors in the situation. The perception of environment depends on the objective conditions, individual difference factors and perception processes. If the subjective perception determines that the environment is within the optimal range of stimulation, the result is equilibrium or homeostasis. In contrast, if the environment is experienced as outside the optimal range of stimulation that one or more of the psychological states like arousal, overload, reactance is likely. The presence of these psychological states is linked with coping strategies. If coping strategies used by individuals are successful, adaptation or adjustment occurs which may be followed by aftereffects like lowered frustration tolerance, fatigue, reduced ability to cope with the next stressors impinging on the individuals immediately. The cumulative aftereffects may also include a degree of learning about coping with future occurrences of undesirable environmental stimulations. In case the coping strategies are not successful arousal and stress will continue. It is possible that the level of stress will be increased largely because the individual may perceive that the coping strategies have failed. In such cases potential aftereffects resulting from inability to cope may be seen which includes learned helplessness, exhaustion, mental disorders, performance deterioration and so forth. Finally, Fischer *et al.* (1984) made provisions of feedback loop in their model, which indicates that individual experiences with the environment influence perception of the environment for future encounters and also contribute to individual differences for future experiences. According to Fischer *et al.* (1979) the model attempts to integrate the various mediating concepts that have been applied to explain man-environmental relationships.

Summary

One major focus of this chapter is to discuss methodological issues so that accurate knowledge about the relationship between environment and behaviour is discovered. The nature and functions of major theories developed in environmental psychology is also discussed so that the links between environment and behaviour is explained.

The ultimate goal of research is development of theories

that explains causality. However, it is essential to be careful about various methodological problems pertaining to validity that poses threat and questions the conclusions drawn. In this regard four different types of validity applicable to design methodologies is presented. The research methods frequently used in environmental psychology are discussed under the experimental and non-experimental research categories. Some major studies conducted under various research paradigms are discussed.

The field of environmental psychology is relatively new as compared to other areas of psychology and thus the field lacks focused direction and sufficient data set on which sound foundation of theories are based. However, some major theories developed across the period of time are discussed and supportive empirical studies are mentioned. Finally, the eclectic model developed by Fischer *et al.* depicting the environment-behaviour relationship is discussed.

REFERENCES

Altman, I. (1975). *The Environment and Social Behavior.* Monterey, California: Brooks/Cole.

Averill, J. (1973). Personal control over aversive stimuli and its relationship to stress. *Psychological Bulletin,* 80, 286-303.

Baird, L.L. (1969). Big school, small school: A critical examination of the hypothesis. *Journal of Educational Psychology.* 60, 253-260.

Barker, R.G. (1960). Ecology and motivation. In M.R. Jones (Ed.), *Nebraska Symposium on Motivation* (Vol. 8). Lincoln: University of Nebraska Press.

Barker, R.G. (1968). *Ecological Psychology: Concepts and Methods for Studying the Environment of Human Behaviour.* Stanford, California: Stanford University Press.

Barker, R.G. (1979). Setting of a professional lifetime. *Journal of Personality and Social Psychology,* 37, 2137-2157.

Barker, R.G., & Gump, P.V. (1964). *Big School, Small School: High School Size and Student Behavior.* Stanford, CA: Stanford University Press.

Baum, A., Greenberg, C.I. (1975). Waiting for a crowd: The behavioral and perceptual effects of anticipated crowding. *Journal of Personality and Social Psychology,* 32, 671-679.

Baum, A., & Paulus, P.B. (1987). Crowding. In D. Stokols & I. Altman (Eds.). *Handbook of Environmental Psychology (Vol. 1).* New York: Wiley.

Baum, A., Singer, J.E., & Baum, C.S. (1981). Stress and the environment. *Journal of Social Issues*, 37, 4-35.

Bell, P.A., & Barnard, S.W. (1977). *Sex Differences in the Effects of Heat and Noise Stress on Personal Space Permeability*. Paper presented at the meeting of the Rocky Mountain Psychological Association, Albuquerque, May.

Brehm, J.W. (1966). *A Theory of Psychological Reactance*. New York: Academic.

Broadbent, D.E. (1971). *Decision and Stress*. New York: Academic Press.

Cohen, S. (1978). Environmental load and the allocation of attention. In A. Baum., J.E. Singer., & S. Valins (Eds.), *Advances in Environmental Psychology* (Vol. 1), N.J.: Erlbaum.

Cohen, S. (1980). Aftereffects of stress on human performance and social behavior: A review of research and theory. *Psychological Bulletin*, 87, 578-604.

Cook, T.D., & Campbell, D.T. (1976). The design and conduct of quasi-experiments and true experiment in field settings. In M.D. Dunette (Ed.), *Handbook of Industrial and Organizational Psychology* (pp. 223-316). Chicago: Rand McNally.

Cook, T.D., & Campbell, D.T. (1979). *Quasi-Experimentation: Design and Analysis Issues for Field Settings*. Chicago: Rand McNally.

Cooper, C. (1971). St. Fracis square: Attitudes of its residents. *A.I.A. Journal*. December, 22.

Desor, J.A. (1972). Toward a psychological theory of crowding. *Journal of Personality and Social Psychology*, 21, 79-83.

Donnerstein, E., & Wilson, D.W. (1976). Effects of noise and perceived crowding on ongoing and subsequent aggressive behavior. *Journal of Personality and Social Psychology*, 34, 774-781.

Dubos, R. (1980). *Man Adapting*, 2nd edition. New Haven, CT: Yale University Press.

Duffy, E. (1962). *Activation and Behavior*. New York: Wiley.

Edney, J.J. (1975). Territoriality and control: A field experiment. *Journal of Personality and Social Psychology*, 31, 1108-1115.

Emiley, S.F. (1975). The effects of crowding and interpersonal attraction on affective responses, tasks performance and verbal behavior. *Journal of Social Psychology*, 97, 267-271.

Epstein, Y.M., & Baum, A. (1978). Crowding: Methods of study. In A. Baum & Y. Epstein (Eds.), *Human Responses to Crowding*. Hillsdale, N.J.: Erlbaum.

Evans, G. (1978). Human spatial behavior: The arousal model. In A. Baum & Y. Epstein (Eds.) *Human Response to Crowding* (pp. 283-303). Hillsdale, NJ: Erlbaum.

Evans, G., Cohen, S. (1987). Environmental stress. In D. Stokols and I. Altman (Eds.), *Handbook of Environmental Psychology*, New York: Wiley.

Fischer, J.D., Bell, P., & Baum, A. (1984). *Environmental Psychology* (2nd ed.), New York: Holt, Rinehart & Winston.

Glass, D., & Singer, J. (1972). *Urban Stress*. New York: Academic.

Greenberg, C.I. (1979). Toward an integration of ecological psychology and industrial Psychology: Undermanning theory, organization size and job enrichment. *Environmental Psychology and Nonverbal Behavior*, 3, 228-245.

Hebb, D.O. (1972). *Textbook of Psychology* (3rd ed.). Philadelphia: Saunders.

Helson, H. (1964). *Adaptation-Level Theory: An Experimental and Systematic Approach to Behavior*. New York: Harper.

Hershberger, R.G., & Cass, R.C. (1974). Predicting user responses to buildings. In D. Carson (Ed.), *Man-Environmental Interactions: Evaluations and Applications* (Part 2), Stroudsburg, Pa: Dowden, Hutchinson & Ross.

Jain, U. (1987). *Psychological Consequences of Crowding*. New Delhi: Sage.

Kahneman, D. (1973). *Attention and Effort*. Englewood Cliffs, NJ: Prentice-Hall.

Kaplan, R. (1974). Some methods and strategies in the prediction of preference. In E.H. Zube, J.G. Fabos, & R.O. Brush (Eds.), *Landscape Assessment, Values, Perceptions and Resources*. Stroudsburg, Pa: Dowden, Hutchinson & Ross.

Kasmer, A. (1976). The development of a useful lexicon of environmental descriptors. *Environment and Behavior*, 2, 135-169.

Langer, E.J., & Saegert, S. (1977). Crowding and cognitive control. *Journal of Personality and Social Psychology*, 35, 175-182.

Latane, B., & Darley, J.M. (1970). *The Unresponsive Bystander: Why Doesn't he Help?* New York: Prentice-Hall.

Lazarus, R.S. (1966). *Psychological Stress and the Coping Process*. New York: McGraw-Hill.

Lazarus, R.S., & Cohen, J. (1977). Environmental stress. In J. Wohlwill & I. Altman (Eds.), *Human Behavior and Environment*. (pp. 90-127). New York: Plenum.

Lynch, K. (1960). *The Image of the City*. Cambridge, Mass: MIT Press.

McBurney, D.H. (2002). *Research Methods*. Singapore: Thomson Asia Pvt. Ltd.

Michelson, W., & Reed, P. (1975). A time budget record sheet. In W. Michelson (Ed.), *Behavioral Research Methods in Environmental Design*. Stroudsburg, Pa: Dowden, Hutchinson, & Ross.

Milgram, S. (1970). The experience of living in cities. *Science*, 167, 1461-1468.

Miller, J.G. (1964). *Adjusting to Overloads of Information. Disorders of Communication*, Proceedings of the Conference of the Association for Research in Nervous and Mental Disease, December 7-8, 1962. New York.

Moos, R.H., & Gerst, M.S. (1974). *University Residence Environment Scale*. Palo Alto, Calif: Consulting Psychologists Press.

Morgan, J.J. (1916). The overcoming of distraction and other resistances. *Archives of Psychology*, 35, 24, 1-84.

Nagar, D. (1997). Environmental stress and human behavior. In G. Misra (Ed.) *Psychological Stress and Health* (pp. 238-262). New Delhi: Concept.

Nagar, D., & Pandey, J. (1987). Affect and performance on cognitive task as a function of density and noise. *Journal of Applied Social Psychology*, 17, 147-157.

Osgood, C., Suci, G., & Tannenbaum, P. (1957). *The Measurement of Meaning*. Urbana University of Illinois Press.

Oxley, D., & Barrera, M. (1984). Undermanning theory and the workplace: Implications of setting size for job satisfaction and social support. *Environment and Behavior*, 16, 211-234.

Paulus, P.B., Nagar, D., & Camacho, M. (1991). Environmental and psychological Factors in reactions to apartments and mobile homes. *Journal of Environmental Psychology*. 11, 143-161.

Proshansky, H.M. (1972). Methodology in environmental psychology: Problems and Issues. *Human Factors*, 14, 451-460.

Proshansky, H., Ittelson, W., & Rivlin, L. (1970). Freedom of choice and behavior in a physical setting. In H. Proshansky., W. Ittelson, & L. Rivlin (Eds.), *Environmental Psychology: Man and his Physical Setting* (pp. 173-182). New York: Holt, Rinehart & Winston.

Reilly, F.E. (1970). *Charles Peirce's Theory of Scientific Method*. New York: Fordham University Press.

Rodin, J. (1976). Density, perceived choice and response to controllable and uncontrollable outcomes. *Journal of Experimental Social Psychology*, 12, 564-578.

Rosenthal, R., & Rosnow, R.L. (1984). *Essentials of Behavioral Research: Methods and Data Analysis*. New York: McGraw-Hill Book Company.

Rapoport, A. (1980). *Human Aspects of Urban Form: Towards a Man-Environment Approach to Urban Form and Design*. Elmsford, NJ: Pergamon.

Saegert, S. (1973). Crowding, cognitive overload and behavioral constraint. In W. Preiser (Ed.), *Environmental Design Research* (Vol 2, pp. 254-261). Stroudsberg, PA: Dowden, Hutchinson & Ross.

Scott, W.E. (1966). Activation theory and task design. *Organizational Behavior and Human Performance*, 1, 3-30.

Seligman, M.E. (1975). *Helplessness*. San Francisco: Freeman.

Seta, J.J., Paulus, P.B., & Schkade, J.K. (1976). Effects of group size on proximity under cooperative and competitive conditions. *Journal of Personality and Social Psychology*, 34, 47-53.

Seyle, H. (1956). *The Stress of Life*. New York: McGraw-Hill.

Sherrod, D. (1974). Crowding, perceived control and behavioral aftereffects. *Journal of Applied Social Psychology*, 4, 171-186.

Sommer, R. (1969). *Personal Space*. Englewood Cliffs, N.J.: Prentice Hall.

Stokols, D. (1972). On the distinction between density and crowding. *Psychological Review*, 79, 275-277.

Sundstrom, E. (1984). Physical environment and social behavior. In K. Deaux and L. Wrightsman, *Social Psychology in the Eighties*, 4th edition. (pp. 430-457). Monterey, CA: Brooks/Cole.

Valins, S., & Baum, A. (1973). Residential group size, social interaction and crowding. *Environment and Behavior*, 5, 421-439.

Wicker, A. (1979). *An Introduction to Ecological Psychology*. Monterey, CA: Brooks/Cole.

Wicker, A., & Kauma, C. (1974). Effects of a merger of a small and large organization on members' behaviors and experiences. *Journal of Applied Psychology*, 59, 24-30.

Wicker, A.W., & Kirmeyer, S.L. (1976). From church to laboratory to national park: A programme of research on excess and insufficient populations in behavior settings. In S. Wapner, S.B. Cohen, & B. Kaplan (Eds.), *Experiencing the Environment* (pp. 157-185). New York: Plenum.

Wirth, L. (1938). Urbanization as a way of life. *American Journal of Sociology*, 44, 1-24.

Wohlwill, J.F. (1974). Human response to the levels of environmental stimulation. *Human Ecology*, 2, 127-147.

Wohlwill, J.F. (1974). Environmental aesthetics: The environment as a source of affect. In I. Altman & J.F. Wohlwill (Eds.), *Human behavior and environment: Advances in Theory and Research* (Vol 1). New York: Plenum.

Worchel, S., & Teddlie, C. (1976). The experience of crowding: A two-factor theory. *Journal of Personality and Social Psychology*, 34, 30-40.

Yerkes, R.M., & Dodson, J.D. (1908). The relation of strength of stimulus to rapidity of habit- formation. *Journal of Comparative Neurology and Psychology*, 18, 459-482.

Zlutnick, S., & Altman, I. (1972). Crowding and human behavior. In J.F. Wohlwill & D.H. Carson (Eds.), *Environment and the Social Sciences* (pp. 44-60). Washington, DC: American Psychological Association.

PART 2

PROCESSES OF PERSON-ENVIRONMENT TRANSACTION

3

ENVIRONMENTAL PERCEPTION AND COGNITION

Human behaviour is strongly embedded in the ecological context. We perform a range of activities like chatting, playing, eating and relaxing in the physical environment. In fact the way we think, feel and behave is largely influenced by the physical provision of the environment. For example, the students may feel that they are not able to perform well because their hostel is located close to the busy and noisy shopping complex or the hostel room is small and the furnishing is not proper. Similarly, some people prefer to enjoy walking along college pathways that are surrounded by natural landscaping with lot of trees, shrubs and greenery. We have also noticed that in summer, people relax and enjoy areas located in hill stations or select a place near the sea beach. The travel environment from home to office may provide pleasant emotional state and satisfaction if the drive is leisurely. In contrast long drive from home to work crossing busy intersections and streets makes as feel uneasy, tired and at times annoying. In other words the physical features of the setting affect our behaviour. The range of activities is largely dependent on our accuracy with which we perceive the various environmental attributes. We have also noticed that even when the actual physical stimulus is identical, the perceptual experiences are different across individuals. Thus, it seems that environmental perception involves the properties of the environment and the characteristics of the person. In other words, some of the explanations regarding differential perceptual experiences lies within the individual and some within the environment. A comprehensive study of environmental perception requires a holistic approach where the individual processes and the components of the environment are not compartmentalized but are viewed as a unit.

Before approaching the concept of environmental perception and its characteristics, it is necessary to briefly discuss the conventional viewpoints of perception as has been used in psychology.

Conventional Approaches to Perception

Before discussing environmental perception a brief discussion on conventional approaches to perception and its associated terminology is required. In formal terms, if any passing source of physical energy activates our sense organ, the energy external to the individual is called stimulus. A stimulus produces a response in the sense organ. When the stimulus is strong enough to activate the receptor cells, which are sensitive to specific form of energy (e.g. light, sound, heat), the process is called sensation. Perception represents a step beyond sensation. The primary difference is that sensation can be thought of as an organism's first encounter with a raw stimulus. In contrast, when mediational processes in the brain integrate various types of sensations so that we recognize, organize, interpret and analyze a pattern out of these sensations, the process is called perception. We will briefly discuss three major conventional traditions to perception. These perspectives are called—*constructivist, structuralism* and *functionalism.*

The *constructivist* perspective emphasizes that perceptions are constructed from sensations and from long-term memory with similar sensations. In other words, perception consists of the information processed on the basis of present sensation and memory of similar sensation. Based on the texture, size, colour and pattern we are able to recognize various objects like table, chairs. The *structuralism* perspective is based on the role of physiological structures in the brain responsible for producing perceptions. This viewpoint is criticized because it fails to incorporate the cognitive processes that are involved in perception. In other words the process involved in perception is based on comparison of our past memory with newly experienced stimuli. Finally, the perspective based on *functionalism* is based on the notion that the perceptual processes are based on the necessity of the organism to get along with the environment. For example, people compare present sensations with the past ones to derive conclusion regarding food, shelter or danger.

Concept, Nature and Characteristics of Environmental Perception

Environmental perception is a unique, complex and dynamic process. Our ability to conduct various activities and affairs of our daily lives is dependent to a large extent on how we perceive our physical surroundings. The individual is not a passive recipient but is largely involved in perceiving the environmental attributes. Understanding the processes involved in environmental perception may enable the architects and planners to design offices, factories to residential areas which are congruent gant with the occupant's needs, desires and aspirations.

The individual is exposed to a variety of stimulus from the environment. The environmental perception deals with the process through which the individual organizes this real world stimulation and derives cohesive, meaningful and integrated picture of the real world settings. In other words the process of apprehending the features of the immediate physical environment through sensory input is called environmental perception. The process involved in environmental perception is closely related to and interacts with other processes, which pertain to environmental cognitions and attitudes. All these three processes constantly influence each other. Human perception is understood as a holistic process. The whole is greater than sum of its parts. The principles of proximity and closure enable the individual to perceive distinctive stimulus in a collective and holistic pattern. The focus of traditional psychological studies on perception was on perceptual processes such as the way people perceive distance, size and movement. Typically, isolated objects rather than perception of large scale or molar physical environment were observed. The work of Ittelson (1976) provides valuable insights into the unique nature of environmental perception. The emphasis of environmental perception is on questions dealing with how people perceive a complex array of many objects or molar environment. For instance, residential complexes, neighbourhood, factories and office settings are subject matter of studies of environmental perception. Perception of isolated objects like distance, size if viewed in meaningful context enhances its perceptual recognitions.

People are embedded and surrounded by the large-scale environment. Thus, perception of surrounding requires constant movement. To examine all aspects of the surrounding adequately and completely multiple perspectives is essential. Environmental perception extends beyond simple observation to exploration from different vantage points. For instance, a person who occupies a new apartment in multistoried residential complex may have to walk through the new setting a number of times to experience its distinctive features like positioning of doors, windows, staircases, elevators, garden, play areas and shopping provisions. Perceptual information about the different aspects of residential setting is gathered through a variety of sense organs like smell, sight and touch. The knowledge gained through perceptual experiences about the new environment is abundant and is comprised of both central and peripheral information. Finally, another unique feature of environmental perception is related to purposive action. When individual encounters the environment he actively explores, sorts and categorizes the inputs, which he receives from his actions in the said environment, which is not random but purposefully directed. We constantly make mental notes of different aspects of the environment that are unique and distinctive so that our future travel in various areas of the city becomes easier.

Adaptation and Perception of Change

Considering time as variable in environmental perception, two important phenomena related to adaptation and change needs mention.

Adaptation

Often we all have noticed that the perceptible stimulus does not change across the period of time. In such situations the response to constant stimulus becomes weaker and the phenomena of adaptation or habituation are seen. Usually, people living near the railway tracks for a prolonged period of time are exposed to loud noise. These people get desensitized to such stimulation that is otherwise annoying to others who live in silent areas. The same desensitizing, adaptation or habituation effect is seen in people who are exposed for a long period of time to other type of stimulations like crowding, foul odour, air pollution and heat.

The terms habituation and adaptation is often used interchangeably. However, sometimes the distinction is made between the two terms. Habituation is more closely associated with physiological process while adaptation effects involve cognitive processes. Physiological explanations for habituation are based on the notion that when people are repeatedly exposed to constant stimulus (e.g. noise, crowding, pollution) the receptors are not able to fire frequently. Cognitive explanation of the phenomena emphasizes a cognitive reappraisal of the stimulus as less important or less threatening. When the individual is exposed to noxious stimulus for the first time he allocates considerable attention and may feel disturbed. However, when the same stimulus is presented on repeated occasions a perceptual shift is seen and the individual reappraises the stimulus and regards it as less threatening. People living near the industries simply learn to respond to bad odour as less intense and non-threatening. One important factor in explaining the process of adaptation relates to the degree of predictability. The results of many studies reveal that adaptation is easy when the stimulus (e.g. noise) is predictable and constant. However, when the noise is irregular and unpredictable the process of adaptation becomes difficult and the individual gets more disturbed (Glass & Singer, 1972).

Perception of Change

Often people who have adapted to environmental stimulation (air pollution) report change in their perception at least in those cases when the intensity of the stimulation shows variation. The perception of change can be explained within the Weber-Fechner framework of psychophysics. The framework explains that when the difference between old intensity of a stimulus is very less even a much smaller increment in the intensity is sufficient to produce change in perception and is detected. In contrast, when the original stimulus is of very high intensity, a much larger increment in stimulus value is required for a change to be noticed. The applicability of Weber-Fechner law in explaining perception of change is seen for all forms of stimulation (e.g. sound, light, pressure). However, it seems that the law should apply not only in laboratory condition to individual stimuli but to urban pollution. In other words, a community living in low pollution area should feel threatened when thick clouds of blown smog are seen suddenly. In contrast, communities residing in polluted areas with heavy smog should require high levels of additional pollution before displaying reactions of annoyance and uneasiness.

Perception of Natural Hazards

A brief discussion regarding cognitive organization of complex perception is being made with special reference to perception and prevention of natural hazards or disasters. The question arises what factors influence our awareness of hazard perception and its potential consequences on being hazard victims? Attempt will be made to discuss three major factors in hazard perception. These are labelled as *crisis effect, levee effect and adaptation* (Fischer, Bell & Baum, 1984).

The *crisis effect* pertains to the fact that perception of hazard is greatest immediately during and following its occurrence. However, the intensity of the effect is substantially reduced between disasters. For example, during severe drought several stringent measures are immediately taken to conserve water and to devise strategies of water harvesting. Effort to conserve water is not taken seriously between droughts and limited attention is given to limit population in areas that are drought prone. The same phenomena holds true for other natural hazards like floods, earthquakes.

The *levee effect* pertains to the fact that soon after the measures taken to deal with the disaster, protective zones are demarked and people are positioned around these areas. For instance, levee are built to protect flood water in the river so as to prevent water from entering the populated areas. People are prone to thinking that the levee is going to protect them from floods. However, levee is built on projected figures that often go wrong.

The *adaptation effect* pertains to the fact that when disasters are frequent people get used to their effect and live amidst disasters. For instance, in earthquake-prone areas in India as well as in other countries people display high level of adaptation so much so that it no longer frightens the inhabitants living therein. People generally discount the possibilities that the earthquakes might victimize them or their families.

A number of factors influence adaptation to potential hazards. For instance, the threshold of adaptation is observed to be high for those people whose well-being is dependent on the hazard-prone area. The business community residing in earthquake-prone areas of Bhavnagar, Gujarat refused to vacate their city even when countless people deserted their city and the frequencies of quakes were severe and very frequent. It seems that people who have survived the dangerous consequences of quakes are more aware of the risks than those who are not having such experiences. Thus, frequency of disasters influences cognitive judgement of potential danger, as does the experience of similar disasters.

Personality factors are also found to be associated with

the perception of hazards. In one study Sims and Baumann (1972) found that most of the tornado related death took place in the South where the concentration of tornadoes was less relative to Midwest where the concentration of tornadoes was heaviest. Even after eliminating the obvious natural explanations such as density of the region and intensity of the storms, the study suggested evidence of regional differences in subjective perception of danger and subsequent level of needed preparation. The Illinois and Alabama residents were interviewed and differences on personality factor related to locus of control were found. The Illinois residents felt luck (external control) has far less to do with their fate (internal control) relative to their counterparts who were from the Alabama state. The Illinois residents appear to take more precautions (e.g. listening to weather forecasts, warning neighbours) when storms approached, whereas, the Alabama residents paid little attention to weather reports. Based on the results it seems that personality factors at least to some extent determine human perceptions regarding control over hazards.

Environmental Assessment

One major issue of environmental perception with regard to environmental assessment is to explore ways of determining the components of scenic quality. Greene (1983) has advocated three broad approaches for assessing the scenic quality. These approaches are Descriptive, Physical-perceptual and Psychological. A brief description of these approaches is given below:

Descriptive Approach

Landscape architects primarily use this approach. The form, colour, and texture are recognized as basic elements of scene. This approach is frequently been used for environmental assessment. The limitation of this approach lies in its subjectivity. This approach heavily relies on expert's opinion and disregards the public views.

Physical-Perceptual Approach

This approach focuses on quantifying the physical properties of the objects in a scene and uses mathematical models. Usually, a panel of judges determines the value of the scene. Hamilton and Schmidt (1969) conducted a study in New York State to assess the individual's preference for landscapes. The study reports a range of factors associated with preferences. These factors relate to the area of immediate vegetation, vegetation multiplied by area of water and so forth. In a similar vein, the scenes of the Connecticut river was assessed and analyzed by Zube, Pilt & Anderson (1974). In this study scenic quality was found to be associated with various components like land-use compatibility, absolute relative relief, height contrast, and density of edges of bodies of water. Other scholars have found that the scenic quality is judged in terms of factors pertaining to debris in the streambed, stream velocity, natural water area, ruggedness, and naturalism (Pitts, 1976; Palmer and Zube, 1976). The physical-perceptual approaches are found to be useful in predicting valid assessment of scenes across cultures as well as consistency within cultures (Zube and Mills, 1976; Anderson, Zube & MacConnel, 1976). This approach is found to be useful when the predictors are simple to work with and are less in numbers. However, this technique becomes cumbersome with the complexity and large number of predictors.

The Psychological Approach

This approach is very similar to the Physical-Perceptual approach. However, instead of using the physical components of the scene, psychological organization of elements embedded in the scenes is used as predictors. For instance, the psychological model incorporates predictors like complexity, ambiguity and coherence. The panel of judges is usually required to rate the beauty and scenic quality. The same or other panel of judges is required to evaluate various specific dimensions of the scene pertaining to complexity, ambiguity, uniqueness and spaciousness.

We perceive the same features of the environment differently. For example, although the objective properties of the stimulus are the same, a contractor, a tourist, and a farmer have very different perception of the forest area. The contractor may look at the forest in terms of the quality of trees that can be cut for financial gains; the tourist may view the overall forest in terms of its beauty. Finally, a farmer living nearby the forest area may perceive the forest in terms of firewood that can be used to cook food. Our attitudes are determined by the different ways we appraise the given environment. Attitudes to the environment are important because of two major reasons. First, attitudes influence our degree of appreciation of the environment in terms of our likes and dislikes and second, they are related to how we use and abuse the environment.

The Kaplan and Kaplan Preference Model

The focus of the Preference model is on factors that tend to influence our likes and dislikes for different types of environment. The major contributors to research in the area on environmental evaluation are Kaplan and others (S. Kaplan, 1975, 1973; R. Kaplan, 1976; Canter, 1968, 1969). The work of Canter attempts to focus on ways in which the meaning we give to our environmental evaluation influences our behaviour. The research ideas can be meaningfully used in the planning and designing the housing projects.

One major issue is related to the question as to why people perceive the same environment differently? The clues for such questions can be traced in linking environmental cognition with environmental evaluation (Kaplan and Kaplan, 1978). We display different emotional reactions in response to different environment. However, our feelings are influenced by our prior experiences. There are considerable individual differences in perceptions of the environment and such differences are reflected in our environmental evaluation. How do we explain the fact that given a choice some people prefer living in crowded areas while others prefer living in remote areas, still other people prefer living in tropical areas, or prefer living in desert areas. To answer such questions we will now focus our discussion on research on environmental preference.

A comprehensive model of environmental preference constructed by Kaplan and Kaplan (1978) will be discussed. These researchers have asked the respondents to classify a large number of slides of various landscapes on various parameters such as likes—dislikes, similar—dissimilar. The researchers had used sophisticated mathematical techniques to identify elements in the scenes that lead to this classification and evaluation. Finally, several factors that tend to predict preferences were identified. Four aspects of the environment that were identified in relation to our emotional experiences were labelled as coherence, legibility, complexity and mystery. A brief description of these factors is given below:

Coherence refers to the degree to which a scene "hangs together". In other words it refers to the degree of organization of parts and how well the whole fits together. High coherence denotes greater degree of liking.

Legibility refers to the degree of distinctiveness or how easily the observer can "read" the environment. Legibility enables the viewer to categorize the contents of the depicted scene. With the increase in the legibility the preference for a given scene is enhanced.

Complexity refers to number and variety of elements in a scene. The greater the complexity the greater the liking.

Mystery refers to the degree to which a scene contains hidden information. The observer is drawn into the scene to search for information. Mystery also promotes liking for the scene.

The model of Kaplan advocates that information content of a scene is important for preference. The elements of the scene that relate to coherence and legibility facilitate the observers to orient themselves within the scene. While the elements of complexity and mystery enhances the observer involvement with the environment. Research in this area seems to suggest that these different dimensions of the environment operate at the individual level through the interaction of physiological and psychological processes.

It seems that a "happy medium" rule is applicable and that optimal level of environmental elements exist which contributes to high degree of appreciation and satisfaction in the observer. In other words a highly coherent, easily legible, simple and well-known environment will be dissatisfying to the observers. Similarly, an environment which lacks coherence, is confusing, overtly complex, and mysterious and strange will also be dissatisfying to the viewers. The focus of one of the study reported by Fischer, Bell and Baum (1984) is on the amount and intensity of information contained in the environment. They have used the information load approach to explain the environmental impact on people. The optimal level of arousal and information load is determined by the individual's past experience. It is possible that one person may like the metropolitan environment because they contain high levels of novelty, mystery and complexity. In contrast, the same reasons may be responsible for another person dislike for the metros. Thus, it is important to identify major dimensions of the environment that will influence the person's affective response. This will allow us to predict the person's evaluation only when individual's life experiences are known.

Environmental Personality

Any scientific approach starts with some sort of categorization or description of the subject matter. Personality theory has always been based on the environment and the traits that are being derived are based on the person's reaction to their environment. Many personality theorists adopt a person-in-context model. A number of environmental psychologists have attempted to describe the environmental personality using questionnaire methods (McKechnie, 1974). His 184-item questionnaire entitled "Environmental Response Inventory" measures eight dispositions towards the environment. These are described as follows:

Pastoralism

The tendency to oppose land development preserve open space, accept natural forces as influences and prefer self-sufficiency.

Urbanism

The tendency to enjoy high density living, and appreciate the varied interpersonal and cultural stimulation found in city life.

Environmental adaptation

The tendency to favour the alteration of the environment to suit human needs and desires, oppose developmental controls, and prefer highly refined settings and objects.

Stimulus seeking

The tendency to be interested in travel and exploration, enjoy complex or intense physical sensations, and has very broad interests.

Environmental trust

The tendency to be secure in the environment, be competent in finding your way around, and be unafraid of new places or being alone.

Antiquarianism

The tendency to enjoy historical places and things, prefer traditional designs, collect more treasured possessions than most other people, and appreciate the products of earlier eras.

Need for privacy

The tendencies to need isolation, not appreciate neighbours, avoid distraction and seek solitude.

Mechanical orientation

The tendency to enjoy technological and mechanical processes enjoys working with hands, and care about how things work (Gifford, 1987; Cassidy, 1997).

The personality approach can be used in designing the

housing projects for persons who move from their original home to the new housing environment (Lee, 1976). The personality perspectives that assume stability of traits are subject to criticism as some scholars advocate that situations may produce similar behavioural pattern irrespective of individual differences. Another group of scholars prefer a cognitive style approach to individual differences. Finally, stable personality trait theories are also criticized by researchers who focus on environmental control models (Cassidy, 1997; Barker & Wright, 1955).

Link between Environmental Cognition, Perception and Evaluation

Many researchers have used the term perception and cognition interchangeably. Perception deals with the direct sensory experiences of a situation or an environment. The term environmental cognition refers to the process by which information is categorized, sorted, organized and structured and placed in meaningful categories. Finally, the process of evaluation goes one step beyond cognition. It refers to the addition of values and preferences to structured cognition. For instance, it is possible for one group of observers to see undesirable features of slum environment, while the other group perceives the same neighbourhood as pleasant and reasonable. Thus, all three processes are involved simultaneously whether the person is engaged in taking small or major decision. People act on the basis of information that is collected (perception), encoded (cognition) and judged (evaluation).

Environmental Cognition and Cognitive Mapping

Environmental cognition is established as a major area of interest to environmental psychology. Studies pertaining to environmental cognition have attracted scholars working in a range of disciplines like geography, economics, sociology, architecture design and regional planning (Fischer, Bell and

Baum, 1984; Lynch, 1960). However, systematic empirical work on environmental cognition is in its preliminary stage (Golledge, 1987). Nevertheless, research studies on environmental cognition is influenced by the ideas and insights gained from psychology (Brunswik, 1944; Ittelson, 1951; Lewin, 1936; Kaplan, 1973; Tolman, 1948).

Perspectives and Theoretical Orientation of Environmental Cognition Research

A brief description of two major themes that has guided the environmental cognition research is given below:

Positivism

The developmental roots of environmental cognition research can be traced in the underlying philosophy of empiricism, positivism and neopositivism. The positivistic approach emphasizes on epistemology and analytic methodology and treats values and beliefs as metaphysical. The positivistic approach is "process oriented" and advocates that the understanding of objective reality is possible only when atomistic facts are ascertained. The role of experimenter/researcher is believed to be passive with respect to the world of objective reality. This approach relies on hypothesis testing, reliability and public verification of results as well as getting the generalization of results across situations (Golledge, 1987). The positivistic approach envisages that the dialectical relationship between man and environment is mediated by cognition. However, some of the assumptions of positivism were subjected to criticism and were abandoned (Couclelis & Golledge, 1983). For instance, the influence of Kantian theory of knowledge is present in the earlier research of environmental cognition (Moore & Golledge, 1976).

Interactional

The creative power of the human mind was acknowledged and the mental reality was considered as the primary object of study. This position questioned the positivistic position that treated the researcher as passive observer of an objective reality. The living interaction between researcher and human subject was emphasized. The interactional perspective advocates that, based on the past experiences and knowledge people perceive and respond to their environment. Following the interactional framework focal issues in environmental cognition research was shifted and fundamental questions regarding the filtering of information through senses, its storage, and access were studied (Moore & Golledge, 1976; Wapner, Kaplan & Cohen, 1973).

The interaction approach attempts to search the situations mediating between individual and features of the environment. For example, it is possible that one person may have a preferred behavioural response to a given context/situation. However, his response is often guided by societal or cultural constraint. For instance, a professional residing in city environment and having its ancestral roots in rural area holds strong belief in inter-caste marriage. But, he is not able to disregard the role of context and societal/group pressure emanating from the context and opts for marriage against his will. In other words, it seems that forces emanating from context/group have changed his behavioural response regarding marriage. Thus, interactional framework advocates that the direct relationship between person and environment is mediated by context/situation. A number of scholars have pointed out that there are significant cultural differences in the way environment and space are generally cognized (Rapoport, 1977; Newell & Simon, 1972).

Concepts and Theories

A variety of theories and concepts have filtered from various disciplines including psychology have influenced the conceptual and theoretical framework of environmental cognition research. A brief summary is presented below:

Role of Development

Research focusing on the development of spatial representation increased in quantity during the 1970s. The work of Piaget and his colleagues (Piaget, Inhelder, 1967; Piaget, Inhelder; & Czeminska, 1960) highlighting the progress of knowledge through different stages was acknowledged by a number of scholars and provided bases for research in diverse environmental settings (Evans, 1980; Hazen, Lockman & Pick, 1978). These scholars have supported the developmental sequences proposed by Piaget and his colleagues that posit an 'egocentric stage', an 'allocentric stage' and a 'geocentric space'. All space related knowledge is referred to as self in the egocentric stage. In the allocentric stage a relative space is constructed that is independent of self. Furthermore, elements of sequences, directionality and some relational concepts are associated with the allocentric stage. Finally, the geocentric stage involves the construction of an absolute space independent of self and specific objects in which the people, places and things are related to one another through spatial principles like closeness, separateness and proximity. With regard to the developmental progress in the use of different spatial elements (e.g. self, landmark, route, frame of reference) sequence some inconsistent results are obtained (Spencer & Darvizeh, 1981).

Some evidence suggests that with regard to reference points people use both naïve and sophisticated ways. Research done by Pick and his associates reveal that an immature way of using landmarks is systematic one-to-one sequencing of several landmarks. In contrast, a sophisticated and mature way is to use a relational space consisting of a series of landmarks as spatial structure (Pick & Lockman, 1981). Furthermore, empirical evidence exist that developmental changes that occur with age are very similar to the changes that occur in experimental settings across time (Golledge, 1987; Evans & Pezdek, 1980). Further evidence by researchers confirming the increase in location and relational accuracy with gain in experience exists. It seems that with increase in experience more information is accumulated. The e..sence of Piaget's work is that the spatial knowledge has its origin in development. This knowledge progresses from different stages from overt and motor to covert and allocentric. Both topological and structural geometric concepts are incorporated into the environmental cognition as development proceeds through the allocentric stage. In other words, spatial understanding is sharpened at different stages of development. The linkages between different places are not only established but also an understanding about the previously unconnected places is also created. A three-phased developmental sequence for environmental cognition was given by Siegal and White (1975). The first phase represents the identification of landmarks followed by progression to route and finally, progressing towards configurational representation. While similar in many respects, the developmental models of environmental cognition described above are different with regard to sequencing of spatial knowledge.

Information Processing

The information processing model utilized by many researchers (R. Kaplan, 1976; S. Kaplan, 1973) in the area of cognitive mapping is stimulated from the earlier work on the networking theory of Hebb (1949; 1963) and theoretical orientation of Gibson (1979; 1966) on environmental perception. For instance, the functionalist nature of information processing was emphasized by S. Kaplan (1973). According to Kaplan (1973) survival in the hostile environment is dependent on various cognitive activities like object recognition, past experience, anticipation of future events, spatial abstraction and generalization. Furthermore, the principles of contiguity, proximity, sequence and representation are very relevant in understanding the dynamics of information processing mechanisms involved in environmental cognition. The perceptual theory of Gibson emphasizes on information processing as a means for extracting, storing and utilizing information about the external world. His concept of physical and social affordance is extended and applied in diverse settings and contexts by many scholars (Stokols & Shumaker, 1981; Evans, 1980).

The importance of central information processing mechanisms (CPM) and knowledge structures in understanding various aspects of environmental cognition is documented in the literature. Attempts were made to develop computer based process models that translate theories of knowledge storage, acquisition and use into symbolic terms so that specific task situations can be programmed using computers (Hayes-Roth, 1977; 1979; Kuipers, 1980). For instance, the position taken by Kahneman (1973) regarding limited information processing capacity at a given point of time was tested both in the real life situation as well in the laboratory situation with the help of computer process models (Golledge, 1987). In the area of cognitive science experts from various disciplines like engineers, computer scientists, psychologists are giving lot of importance to artificial intelligence modelling to understand the complexity of knowledge acquisition and its use. In this regard various concepts of environmental cognition such as knowledge structure, declarative and procedural knowledge, spatial knowledge, semantic nets, frames are studied (Kuipers, 1982; Minski, 1975; Schank & Abelson, 1977). Furthermore, the concepts discussed above are linked into computer process model (CPM). In one study Schank & Abelson (1977) have examined the relevance of CPM in the area of environmental cognition. It seems that an individual's knowledge about the environment is a function of accumulated and organized perceptual experiences that represents interaction between the person and aspects of the environment. Thus, the knowledge acquisition process is based on episodes from which a more generalized and context independent knowledge structure emerges that tends to sustain a range of spatial behaviour. In short, informational process theories have proven to be useful in examining the selection, storage, reconstruction and use of cognitive information. Thus far, no definite answer regarding various aspects of processing of spatial information is available. However, these areas have generated substantial research interest among interdisciplinary scholars.

Cognitive Mapping and Cognitive Maps

Research studies are conducted in diverse areas of environmental cognition. It is difficult to review all areas of environmental cognition. An attempt will be in this section to review studies conducted in the broad areas of cognitive mapping and cognitive maps. The concept of cognitive maps has been documented in the literature in the earlier phases of the last century (Trowbridge, 1948; Tolman, 1932). However, cross-disciplinary attention on the notion of cognitive maps can be seen in studies done by scholars such as Lynch (1960), De Longe (1962), Appleyard (1970), Stea (1969) and Downs (1970). The early studies focused on demonstrating the difference between images of reality and objective construction of reality. Furthermore, emphasis was given to the analysis of cognitive maps as a problem in cognitive structure *per se*. For the last three decades or so many focused researches are conducted by scholars on cognitive maps and mapping that dealt with complex and varied issues.

Cognitive mapping is concerned with describing and understanding people's awareness of the physical properties of their surrounding. The process of cognitive mapping is a means of structuring, making sense of and coping with the complex external environment. We not only deal with observable physical environment but we are also subjected to various aspects of social and cultural environment that impinge on us on routine basis (Stokols & Shumaker, 1981). A cognitive map is an individual's viewpoint and internal representation of the world.

The comprehensive work of Lynch (1960) amply focuses on human cognitive maps of cities. His work provides a framework for thinking about environmental cognition scientifically by providing research methodology of studying mental maps. One very common practice is to recover the critical cue structure of cities by the process of simply asking people to list places well known to them and then compiling composite scores based on the frequency with which particular places were mentioned. In the field of psychology the concept of cognitive mapping was used by Tolman (1948) in his famous study dealing with how the rats learn to "map" the environment of an experimental maze. To solve our location and relational problems we use our sense organs and search the inputs from the environment. Our long-term memory provides us some cues about the location. Our mental representation or cognition about the layout is termed as cognitive map and the general way of thinking about, recognizing and organizing this layout is called environmental cognition (Fischer, Bell & Baum, 1984). A cognitive map is a representation of the physical environment. The vision as well as sensory and motor modalities is simultaneously involved, making the process of cognitive representation complex. The cognitive maps are stored as complex structures in the brain that may be restructured with respect to the users need. Researchers interested in the study on environmental cognition seeks to address a range of questions pertaining to the ability of storing, recovering, using and accessing information about the environment.

Components of Cognitive Maps

Lynch (1960) has reported five categories to describe and analyze cognitive maps of cities. These categories are labelled as paths, edges, district, nodes, and landmarks. *Paths* serve as the predominant and basic element of a map. They serve as movement channels such as highways, traffic arteries, and bus or railroad lines for people. *Edges* are separate linear elements. They are limits of the maps that tend to follow a line. Unlike paths, edges are not connectors. Physical barriers that separate two districts may serve as edges. Similarly, psychological edges may make two areas distinct. For instance, seashore falling in between two separate rural community areas may serve as edges. A district is a distinctively large physical region or section of a city that is memorable in certain ways and is recognized as having a common character. *Districts* may be ethnically distinctive or functionally distinctive. Usually, district have peculiar characteristic, which help people recognize that they have entered a given district. Some examples of districts are downtown of major cities, boy's hostels in the university campus, police lines and so forth. Nodes are major point of focus. In cities any distinctive or strategic point can serve as nodes. Usually, activities are centred on nodes. Some common nodes are major traffic intersection, railway or bus station, highly visible shopping complexes and so forth. Finally, landmarks are points, which have some common elements of magnanimity. They are specific structures that people use for reference points, which are usually visible from long distance. Sometimes a landmark like Taj Mahal of Agra, Kutub Minar of Delhi, symbolizes the entire city. Landmarks are often found on important nodes. They are used as reference points for indicating distance and direction.

A simple model of cognitive mapping incorporates four elements: an actor, an external situation or environment, a set of outputs from situation to actor called environmental cognition, and a set of output from actor to situation called environment response behaviour. This model posits that spatial behaviour cannot be understood completely in terms of stimulus and response. It is required that focus on an internal mechanism that reorders, recodes and transforms the stimuli in a novel and creative way is needed to get a comprehensive knowledge of environmental cognition. One elementary question asked in cognitive mapping is that when engaged in environmental behaviour what kinds of information are coded, stored and decoded? According to Golledge (1987) one possible framework is to regard the cognitive map as a set of stored propositions about the environment, each having been assigned a truth-value or salience by the actor. In other words the process of cognitive mapping consists of a set of operation designed to code environmental information and subsequently decode it to allow spatial behaviour to take place. The most fundamental skill involved in this transformation process are the acquisition of shape constancy, size constancy and a generalized pattern recognition ability related to concept formation and stimulus generalization (Gibson, 1966).

Functions and Properties of Cognitive Maps

One very useful function served by cognitive map is the efficient movement of people in the environment and locating range of places like shopping, entertainment with accuracy. Without the map our search of desired places of interest might have been random, haphazard and inefficient. In our memory the layout of the places are systematically organized which facilitates us to reach our familiar and even unfamiliar destination points in a very economical manner. Thus, solving our spatial problems is an important function of cognitive maps. Another important function of cognitive map is related to communication of location to people. We are able to visualize directions given by others. In fact, shared environmental representations provide essential symbols and collective memories for shared communication. Often we are able to associate the directions with various components of the cognitive maps like familiar landmarks, paths, and districts.

There are three interrelated properties of cognitive maps: *places, spatial relations and travel plans* (Garling, Lindberg, & Mantyla,1983). Some of the common properties of places are spatial scale, names, perceptual characteristics and function, and psychological attributes like attractiveness, affectiveness and so forth. Other salient characteristics are location, distance, direction and orientation with respect to other places. To mark places specific verbal labels and names are attached to places. A number of studies have delineated important attributes of places. Spatial relations are characteristics related to two or more places. The spatial relations are comprised of proximity relations (e.g. among pairs of points), metric relations with respect to distance and direction and topological relations (e.g. inclusion, ordinal). Some studies have dealt with the question whether or not spatial knowledge can be represented metrically (Baird, Wagner, & Noma, 1981)? Some scholars have noted that metric properties exist among more familiar points, lines, nodes, paths and areas contained within the cognitive maps (Golledge *et al.*, 1982). It seems that the lack of a formal geometry in a segment of cognitive map may be a function of many factors like the degree of knowledge acquisition, the quality of the information acquired (Garling, Book, Lindberg, and Nilsson, 1981).

Factors Influencing Cognitive Maps

One of the major factor influencing cognitive maps is familiarity. A number of studies have reported that more familiar we are with the environment, the more accurate and detailed our cognitive maps will be (Appleyard, 1970; 1976; Evans, 1980). Familiarity with elements of the environment that are adjacent to activity space is uniformly highest. Most frequently the activity space contains the major anchor points of one's cognitive mapping of an environment. Thus individual cognitive maps may be most accurate in widely disparate spaces. A change in emphasis from local to distant areas to anchor a cognitive map is influenced to a large extent by factors related to maturation, aging, and passing through successive life cycles (Golledge, 1987). A number of studies have reported positive relationship between length of residence and environmental familiarity. This relationship is influenced by a variety of psychological and geographical features (Golledge, 1978; Milgram, 1970; Moore, 1979). It seems that individuals have learned to discriminate efficiently those physical environmental objects that are salient to them. When the cognized environment reaches some stable state the environmental transactions becomes more predictable and less stressful with regard to decision-making. It is plausible to expect that all information that passes through perceptual filters be not recalled overtime and that some portion of the information is not recalled properly because people try to cope up with informational overload. Furthermore, factors pertaining to maturation or changes occurring in one's lifestyle tend to change the previous emphasis with regard to different aspects of cognitive mapping (Evans, Brennan, Skorpanich & Held, 1984).

A number of studies have also reported evidence that very young and very old people assign the most important material and symbolic significance to their proximate environment including their home and immediate neighbourhood (Rowles & Ohta, 1983). With regard to gender inconsistent results are found with some studies demonstrating males as compared to females draw more comprehensive sketches of cognitive maps (Appleyard, 1979) while other studies reported no sex differences (Maurer & Baxter, 1972). Finally, high socio-economic status groups draw more accurate maps relative to their low socio-economic status counterparts (Orleans, 1973). It seems that people belonging to high status probable are more mobile and have more experience with the broader areas of city relative to low economic status people.

Errors in Cognitive Maps

Cognitive maps are rough conceptualization and are not very perfect representation of the physical environment. Three major sources of errors are seen in cognitive maps (Evans, 1980; Milgram, 1970). First, we very frequently omit major landmarks and districts besides ignoring minor paths and districts. Thus, maps are often incomplete. The second common error is regarding distortion. We tend to place various features of the physical environment either too close together or too far apart. Problems pertaining to improper alignment of various details are also very frequent. Stevens and Coupe (1978) has pointed out that distortion may occur for distance judgements either along a single route or within a larger configuration. Finally, augmentation of details that are not present in the physical environment figures out in cognitive maps. Often various features are added to the cognitive map that are subjectively created by people and are not present in the environment.

Methodological Issues

A comprehensive list of tasks used and methodology adopted to study various issues related to environmental cognition is comprehensively presented by Golledge (1987). The methodology employed ranged from introspection (Binet, 1894), maze learning (Brown, 1932; Tolman, 1948), sketch map technique (Lynch, 1960), topological clustering and mapping categorization (Appleyard, 1969), imaging scenes from different perspectives (Piaget & Inhelder, 1967), recognizing city scenes (Milgram, Greenwald, Keesler, Waters, 1972) to multidimensional scaling of paired interpoint distance judgement (Golledge & Colleagues, 1969) and use of linear scaling of interpoint distance judgment (Briggs, 1972). Baird (1979) required subjects to interact with a computer in order to develop configuration of places. Other methods include the constructive use of models and toy play with children (Blaut & Stea, 1971; Siegal & Schadler, 1977) and the use of two or three-dimensional models with normal and borderline retarded adults (Golledge, Parnicky & Rayner, 1980; Richardson, 1976).

Attempts were made to deal with some of the problems associated with sketch mapping procedure. The psychologists and geographers utilized a variety of sophisticated methodologies to analyze spatial data recovered from the knowledge structure of individual with considerable success in terms of drawing spatial properties of the cognitive maps (Baird, 1979; Golledge, 1977). However, evidence exists that some part of the cognitive maps are usually spatially distorted with respect to familiarity of the situation, nature of task situation and the degree of interaction between the individual and the environment. It is suggested that the use of converging methods may prove helpful in not only understanding the processes involved in environmental cognition but also in overcoming the weaknesses inherent in separate methodologies (Evans, 1980; Golledge & Rushton, 1976).

A variety of approaches and analytical techniques may be used to externalize configuration that can completely represent internal representation of the location of environmental attributes and its environmental provisions. For any given problem situation, suitable methodology is selected by the researcher that is appropriate with the type of the task and expected outcomes. Sufficient evidence exist that controlled experiments may provide valid and reliable results provided appropriate methodological approach is adopted. The use of scientific approach and appropriate statistical techniques may provide pertinent information about properties of mapping. A plan to design similar experiments and converging methodologies more authentic inferences about cognitive mapping can be derived and better generalization about the environmental cognitive processes is possible. As has been popular in ethological and anthropological tradition (Hall, 1966) more emphasis is given to naturalistic and field research. Future research may be planned keeping in mind methodological diversity is probably more suitable for examining spatial-physical aspects of the environment in naturalistic settings.

Recent Trends in Environmental Cognitive Research

In the field of environmental psychology perceptual processes occupies central position for explaining person-environmental transactions and gaining adequate knowledge about various aspects of environment behaviour. Researchers belonging to various disciplines like geography, sociology, town planning, and anthropology have focused their attention on the ways individuals relate and respond to the physical properties of the environment through perception (Kates & Wohlwill, 1966; Lynch, 1960; Romney & D'Andrade, 1964; 1961; Orleans, 1973). However, there has been lot of variation at the conceptual levels regarding the use of the terms perception and cognition (Downs & Stea, 1973). The prevalent theories of psychology basically focuses on the perceptual responses people give to their spatial and physical characteristic of the environment. In other words, the environmental perception has been studied in terms of the information provided by the formal organization of physical space as a function of cognitive maps. The focus of research is on questions pertaining to the functioning of acquisition, organization and use of spatial information as well as the emotional components involved in these processes. During the last three decades or so a shift is visible in the focus of research with emphasis on the relationship between knowledge and action in studies dealing with environmental perception and cognition.

Another significant trend is shift of interest from exclusively formal aspects of the physical space to the content components through which space is defined. Furthermore, the structuring of environmental knowledge, which was previously centred, on evaluative component is expanded with an inclusion of cognitive and emotional components. For instance, according to Ittelson, Proshansky, Rivlin and Winkes (1974) persons are always part of the situation and environmental perception is mainly a social phenomenon. He has emphasized the multidimensionality of the environment within which they distinguish between physical and social dimension as symbolic meanings and motivational messages as properties of the environment. The general framework of environmental perception has treated stimulus as the "source of information". For the long period of time studies of psychology formulated wrong assumption that two distinct and separate system involving perceptual and cognitive processes were linked by fixed and unidirectional bonds. In contrast, recent trend of research treats the entire perceptual-cognitive system as part of a larger system, whose primary function is the elaboration of information. Thus, the term cognitive map implies a modality of study that treats perceptual and cognitive processes as one entity and not as distinct components.

Another visible trend in studies of spatial cognition is regarding emphasis on the contents of the space and not on the formal characteristics of the physical space. According to Garling and Golledge (1987), an understanding of how people mentally construct cognitive maps, consideration of travel plans they propose and carry out is crucial. Thus, the degree of familiarity acquired in a given environment by the individual that constitute the socio-physical context of the individual's daily life becomes important. In addition, the spatial dimension is treated as a component within which environmental elements are placed and their specificity is marked in terms of perceptual, symbolic and functional representation. In this continuation, the cognitive map is meant to represent three interrelated components: information regarding elements constituting environmental space, the spatial relations which can be established between them in terms of inclusion, direction-distance and proximity, but with respect to references which have a particular perceptual and symbolic salience for people, and travel plans, which include the ways they can reach others starting from a determined point in environmental space (Golledge, Rayner & Rivizzigno, 1982). Furthermore, later studies in the area of spatial cognition tend to analyze the processes of environmental knowledge in connection with goals and plans of action people propose to accomplish in the given environment. Recent research on spatial cognition attempts to incorporate both formal aspects and content of the environmental space represented in a cognitive map by considering relationship between knowledge and action.

The other thrust area of environmental cognition research is to treat the knowledge of place not essentially in terms of spatial properties (e.g. locational, relational, and configurational) but also in non-spatial terms largely because places may be culturally coded with respect to symbolism, values, beliefs (Appleyard, 1979; Stokols & Shumaker, 1981). A number of scholars have noted that various aspects of the environment are provided with symbolic meaning. For instance, Saegert and Winkel (1990) point to a symbolization of home by many women in terms of importance to self, family and social relationships. Home for men symbolizes childhood memories. Finally, cross-cultural work on environmental cognition has focused on various aspects including way-finding activities, visual spatial memory, skill acquisition and so forth.

Summary

Human behaviour is embedded in the ecological context. We have noticed that even when the actual physical stimulus is identical, the perceptual experiences are different across individuals. It seems that environmental perception involves the properties of the environment and the characteristics of the person. The process of apprehending the features of the environment through sensory input is called environmental perception. The process involved in environmental perception is closely related to and interacts with other processes that pertain to environmental cognition and attitudes.

The chapter addresses various questions related to how people perceive a complex array of many objects or molar environment. Multiple perspectives are required to understand all aspects of the surrounding adequately. Considering time as variable two major phenomenon involved in environmental perception related to adaptation and perception of change are discussed. A distinction is made between two terms adaptation and habituation. Often both these terms are used interchangeably. Habituation is more closely associated with the physiological process, while adaptation effects involve cognitive processes. Factors affecting adaptation process are discussed. Studies are also discussed which tends to examine issues pertaining to change of perception even after people adapt to environmental stimulation. A brief discussion is been made regarding cognitive organization of complex perception with special reference to natural hazards and disasters. Three major factors affecting hazard perception labelled as crisis effects, levee effect and adaptation are discussed. In addition, a number of studies have revealed that personality factors are also associated with perception of hazards.

One major issue of environmental perception with regard to environmental assessment is to explore ways of determining the components of the scenic quality. Three major approaches—descriptive, physical-perceptual and psychological are discussed. The strength and limitations of these approaches are also noted. A number of questions like why people perceive the same environment differently? The clues for such question can be traced in linking environmental cognition with environmental evaluation. The Kaplan and Kaplan Environmental Preference model have identified four factors in relation to our emotional experiences which they labelled as: coherence, legibility, complexity and mystery. The model advocated that information content of a scene is important for preference. Finally, environmental personality approach is discussed and its implication in designing housing projects is briefly discussed.

The person engaged in any type of decision-making uses all the three processes—environmental cognition, perception and evaluation simultaneously. Generally, people act on the basis of information that is collected (perception), encoded (cognition) and judged (evaluation). Studies on environmental cognition have attracted attention of scholars working in various disciplines. The area of environmental cognition is a major area of interest to environmental psychology. A brief descriptions of two major themes—positivism and interactional, that has guided the environmental cognition research is presented. A variety of theories and concepts that have filtered from other disciplines have influenced the conceptual and theoretical framework in various areas of environmental cognition is examined. For instance, research of Piaget focusing on the development of spatial representation is discussed. Some evidence exist that developmental changes that occur with age are very similar to the changes that occur in experimental settings. However, some inconsistent results are obtained with regard to the developmental sequence in the use of different spatial elements. Furthermore, the influence of central information processing mechanisms and knowledge structures in understanding various aspects of environmental cognition is reviewed. The information process theories have proven to be useful in examining the selection, storage, reconstruction and use of cognitive information.

While studies are conducted in diverse areas of environmental cognition, an attempt has been made to review studies on cognitive mapping and cognitive maps. Cognitive mapping is concerned with describing and understanding people's awareness of the physical properties of the surroundings. Evidence exists that we not only deal with observable physical environment but we are also influenced by various social cultural environment that impinge on us on routine basis. A cognitive map is an individual's viewpoint and internal representation of the world. The components, functions and properties of cognitive maps are discussed. Some of the major factors influencing cognitive maps are detailed. A range of methodologies adopted to study various issues related to environmental cognition are examined. Recent trends of environmental cognition research are discussed. Although, after the pioneering research done by Lynch, substantial research is carried out in various areas of environmental cognition. However, many aspects of environmental cognition are still in its infancy stage. More naturalistic and field studies are needed to verify and validate the results obtained in the experimental settings. Appropriate methodology is not developed for data collection pertaining to various aspects of environmental cognition. The existing methodological approach fails to comprehensively address issues related to how external representation of subjective information is adequately examined and its properties be analyzed.

REFERENCES

Appleyard, D. (1970). Styles and methods of structuring a city. *Environment and Behavior,* 2, 100-118.

Appleyard, D. (1976). *Planning a Pluralist City.* Cambridge, MA: MIT Press.

Appleyard, D. (1979). The Environment as a Social Symbol. *Journal of the American Association,* 45, 143-153.

Baird, J.C. (1979). Studies of the cognitive representation of spatial relations: An overview. *Journal of Experimental Psychology General,* 108, 90-91.

Baird, J.C., Wagner, M. & Noma, E. (1981). Impossible cognitive spaces. *Geographical Analysis,* 14, 204-216.

Barker, R.G. & Wright, H. (1955). *Midwest and its Children.* New York: Row and Peterson.

Binet, M.A. (1894). Reverse illusions of orientation. *Psychological Review,* 1, 337-350.

Blaut, J.M. & Stea, D. (1971). Studies of geographic learning. *Annals of the Association of American Geographers,* 61, 387-393.

Briggs, R. (1972). *Cognitive Distance in Urban Space.* Unpublished doctoral dissertation, Ohio State University, Columbus.

Brown, W. (1932). Spatial integrations in a human maze. *University of California Publications in Psychology*, 5, 123-134.

Brunswik, E. (1944). Distal focusing on perception. *Psychological Monographs*, 56, 1-48.

Canter, D. (1968). *The Measurement and Meaning and Architecture.* Unpublished Manuscript. Glascow Building Performance Research Unit.

Canter, D. (1969). A intergroup comparison of connotative dimensions. *Environment and Behaviour*, 1, 37-48.

Canter, D. (1983). The purposive evaluation of places: A facet approach. *Environment and Behavior*, 15, 659-698.

Cassidy, T. (1997). *Environmental Psychology: Behaviour and Experience in Context.* Psychology Press, publishers, UK.

Couclelis, H., & Golledge, R.G. (1983). Analytic research, positivism, and behavioral Geography. *Annals of the Association of American Geographers*, 73, 331-339.

De Jonge, D. (1962). Images of Urban Areas: Their Structure and Psychological Foundations. *Journal of American Institute of Planners*, 28, 266-276.

Downs, R.M. (1970). The cognitive structure of an urban shopping centre. *Environment and Behavior*, 2, 13-39.

Downs, R.M., & Stea, D. (Eds.) (1973). *Images and Environment: Cognitive Mapping and Spatial Behavior*, Chicago: Aldine.

Evans, G. (1980). Environmental cognition. *Psychological Bulletin*, 88, 259-287.

Evans, G., Brennan, P., Skorpanich, M.A., & Held, D. (1984). Cognitive mapping and elderly adults: Verbal and location memory for urban landmarks. *Journal of Gerontology*, 39, 452-457.

Evans, G., & Pezdek, K. (1980). Cognitive mapping: Knowledge of real world distance and location information. *Journal of Experimental Psychology: Human Learning and Memory*, 6, 13-24.

Garling, T., Book, A., Lindberg, , E., & Nilsson, T. (1981). Memory for the spatial layout of the everyday physical environment: Factors affecting the rate of acquisition. *Journal of Experimental Psychology*, 1, 263-277.

Garling, T., & Golledge, R.G. (1989). Environmental perception and cognition. In H. Zube and G.T. Moore (Eds.), *Advances in Environmental Behavior and Design* (Vol. 2, pp. 203-236), New York: Plenum.

Garling, T., Lindberg, E., & Mantyla, T. (1983). Orientations in buildings: Effects of familiarity, visual access, and orientation aids. *Journal of Applied Psychology*, 68, 177-186.

Gibson, J.J. (1966). *The Senses Considered as Perceptual Systems.* Boston: Houghton Miffin.

Gibson, J.J. (1979). *The Ecological Approach to Visual Perception.* Boston: Houghton Miffin.

Gifford, R. (1987). *Environmental Psychology: Principles and Practice.* Boston: Allen & Bacon.

Glass, D.C., & Singer, J.E. (1972). *Urban Stress*. New York: Academic Press.

Golledge, R.G. (1977). Multidimensional analysis in the study of environmental behavior and environmental design. In I. Altman and J.F. Wohlwill (Eds.), *Human Behavior and Environment: Advances in Theory and Research* (Vol. 2, pp. 1-42). New York: Plenum.

Golledge, R.G. (1987). Environmental Cognition. In D. Stokols & I. Altman (Eds.), *Handbook of Environmental Psychology*. (Vol. 1, pp 131-174), New York: John Wiley & Sons.

Golledge, R.G., Briggs, R., & Demko, D. (1969). The configuration of distances in intra-Urban space. *Proceedings of the Association of American Geographers*, 1, 60-65.

Golledge, R.G., Parnicky, J.J.,& Rayner, J.N. (1979). *The Spatial Competence of Selected Populations* (Vols. 1&2). Columbus: Ohio State University Research Foundation.

Golledge, R.G., Rayner, J.N., & Rivizzigno, V.L. (1982). Comparing objective and cognitive representations of environmental cues. In R.G. Golledge and J.N. Rayner (Eds.), *Proximity and Preference: Problems in the Multidimensional Analysis of Large Data Sets* (pp. 233-266). Minneapolis: University of Minnesota Press.

Golledge, R.G., & Rushton, G. (1976). *Spatial Choice and Spatial Behavior*. Columbus: Ohio State University Press.

Hall, E.T. (1966). *The Hidden Dimension*. New York: Doubleday.

Hayes-Roth, B. (1977). Evolution of cognitive structures and processes. *Psychological Review*, 84, 260-278.

Hayes-Roth, B. (1979). Distinguishing theories of representation: A critique of Anderson's "arguments concerning mental imagery." *Psychological Review*, 86, 376-382.

Hazen, N.L., Lockman, J.J., & Pick, H.L., Jr. (1978). The development of children's representations of large-scale environments. *Child Development*, 4, 71-115.

Hebb, D.O. (1949). *The Organization of Behavior*. New York: Wiley.

Hill-Lynch, K. (1960). *The Image of the City*. Cambridge, MA: MIT Press, Cambridge.

Ittelson, W.H. (1951). The consistency in perceptual theory. *Psychological Review*, 58, 285-294.

Ittelson, W.H., Proshansky, H.M., Rivlin, A., & Winkel, G. (1974). *An Introduction to environmental psychology*, New York: Holt, Rinchart & Winston.

Kahneman, D. (1973). *Attention Effort*. Englewood Cliffs, NJ: Prentice Hall.

Kaplan, R. (1976). Way-finding in the natural environment. In G.T. Moore & R.G. Golledge (Eds.), *Environmental Knowing* (pp. 46-57). Stroudsburg, PA: Dowden, Hutchinson & Ross.

Kaplan, S. (1973). Cognitive maps in perception and thought. In R. Downs & D. Stea (Eds.), *Image and Environment: Cognitive Mapping and Spatial Behaviour*, Chicago: Aldine.

Kaplan, S.-(1975). An informal model for the prediction of preference. In E.H. Zube, R.O. Brush., & J.G. Fabos (Eds.), *Landscape Assessment*. Stroudsburg, PSW-35. Berkeley, CA,

Kaplan, S. (1979). Perception and landscape: Conceptions and Misconceptions. In G. Elsner & R. Smardon, *USDA Forest Service Report PSW-35*, Berkeley, CA.

Kaplan, S., & Kaplan, R. (1978). *Humanscape: Environments for people*. North Scituate, MA: Duxbury.

Kaplan, S., & Kaplan, R. (1982). *Cognition and Environment: Functioning in an Uncertain World*. New York: Praeger.

Kates, R.W., & Wohlwill, J.F.(Eds.) (1966). Man's response to the physical environment. *Journal of Social Issues*, 22, 1-140.

Kuipers, B. (1980). A model of the acquisition of spatial knowledge. In L. Bowl (Ed.), *Models of Dialogue: Theory and Application*. London: McMillan.

Kuipers, B. (1982). The 'map in the head' metaphor. *Environment and Behavior*, 4, 202-220.

Lee, T. (1976). *Psychology and the Environment*. London: Methuen.

Lewin, K. (1936). *Principles of Topological Psychology*. New York: McGraw Hill.

Lynch, K. (1960). *The Image of the City*. Cambridge, MA: MIT Press, Cambridge.

Maurer, R., & Baxter, J. (1972). Images of neighbourhood among black, Anglo and Mexican-American children. *Environment and Behavior*, 4, 351-388.

McKechnie, G.E. (1974). *ERI Manual: Environmental Response Inventory*. Berkeley, CA: Consulting Psychologists Press.

Mehrabrian, A., & Russell, J.A. (1974). *An Approach to Environmental Psychology*. Cambridge, MA: MIT Press.

Milgram, S. (1970). The experience of living in cities. *Science*, 167, 1461-1468.

Milgram, S., Greenwald, J., Keesler, W., McKenna, W., & Waters, J. (1972). A psychological map of New York city. *American Scientist*, 60, 194-200.

Minski, M.A. (1975). A framework for representing knowledge. In P.H. Winston (Ed.), *The Psychology of Computer Vision*. New York: McGraw-Hill.

Moore, G.T. (1979). Knowing about environmental knowing: The current state of theory and research on environmental cognition. *Environment and Behavior*, 11, 33-70.

Moore, G.T., & Golledge, R.G. (1976). *Environmental Knowing: Theories, Research and Methods*. Stroudsburg, PA: Dowden, Hutchinson & Ross.

Newell, A., & Simon, H.A. (1972). *Human Problem Solving*. Englewood Cliffs, NJ: Prentice-Hall.

Orleans, P. (1973). Differential cognition of urban residents: Effects of social scale on mapping. In R.M. Downs & D. Stea (Eds.), *Images and*

Environment: Cognitive Mapping and Spatial Behavior (pp. 115-130). Chicago: Aldine.

Piaget, J., & Inhelder, B. (1967). *The Child's Conception of Space.* New York: Norton..

Piaget, J., Inhelder, B., & Czeminska, A. (1960). *The Child's Conception of Geometry*, New York: Basic.

Pick, H.L. Jr., & Lockman, J.J. (1981). From frames of reference to spatial representations. In L.S. Liben, A.H. Patterson & N. Newcombe. *Spatial Representation and Behavior Across the Lifespan: Theory and Application.* New York: Academic.

Rapoport, A. (1977). *Human Aspects of Urban Form.* New York: Pargamon.

Richardson, G.D. (1982). *Spatial Cognition.* Doctoral dissertation, University of California, Santa Barbara.

Romney, A.K., & D'Andrade, R.G. (1964). Transcultural studies in cognition. *American Anthropologists*, 66, 140-170.

Rowles, G., & Ohta, R. (1983). *Aging and Milieux.* New York: Academic.

Saegert, S., & Winkel, G. (1990). Environmental Psychology. *Annual Review of Psychology*, 41, 441-477.

Schank, R., & Abelson, R. (1977). *Script, Plans, Goals, and Understanding.* Hillsdale, NJ: Erlbaum.

Siegal, A.W., & Schadler, M. (1977). Young children's cognitive maps of their classrooms. *Child Development*, 48, 388-394.

Siegal, A.W., & White, S.H. (1975). The development of spatial representations of Large-scale environments. In H.W. Reese (Ed.), *Advances in Child Development and Behavior* (Vol. 10), New York: Academic.

Spencer, C., & Darvizeh, C. (1981). The case for developing a cognitive psychology that does not underestimate the abilities of young children. *Journal of Environmental Psychology*, 1, 21-31.

Stea, D. (1969). The measurements of mental maps: An experimental model for studying conceptual spaces. In K.R. Cox & R.G. Golledge (Eds.), *Behavioral Problems in Geography: A Symposium.* Evanston, IL: North-Western University.

Stevens, A., & Coupe, P. (1978). Distortions in judged spatial relations. *Cognitive Psychology*, 10, 422-437.

Stokols, D., & Shumaker, S.A. (1981). People in places: A transactional view of settings. In J.H. Haevery (Ed.), *Cognition, Social Behavior, and the Environment.* Hillsdale: NJ: Erlbaum.

Tolman, E.C. (1948). *Purposive Behavior in Animals and Men.* New York: Appleton-Century-Crofts.

Trowbridge, C.C. (1948). *Comparative Psychology of Mental Development.* New York: International Universities Press.

Wapner, S., Kaplan, B., & Cohen, S.B. (1973). An organismic-developmental for understanding transactions of man in environment. *Environment and Behavior*, 5, 255-289.

White, G. (1945). *Human Adjustments to Floods* (Research paper No. 29), Chicago: University of Chicago, Department of Geography.

4

HUMAN SPATIAL BEHAVIOUR

PERSONAL SPACE AND TERRITORIALITY

With more than thousand studies reported on personal space (Hayduck, 1985) the expanded area of human spatial behaviour has spurred considerable attention from scholars interested in the area of environment and behaviour. Research on human spatial behaviour burgeoned from the popular work of Calhoun (1962), Ardey (1966) and Lorenz (1966). However, it was only after the appearance of two popular books entitled *The hidden dimension* (Hall, 1966) and *Personal space: The behavioural basis of design* (Sommer, 1969) that research interest among scholars sparked and more richness and diversity to the domain of human spatial behaviour was visible. Psychologists have explored the issue of human spatial behaviour under three main areas: personal space, territoriality, crowding. These three areas for research are very much interrelated and interdependent. In this chapter we will explore the issues pertaining to personal space and territorial behaviour in detail and in the next chapter the issues of crowding will be discussed.

Difference Between Personal Space and Territorial Behaviour

The concepts of personal space and territorial behaviour are interrelated. However, certain conceptual and methodological differences between these two concepts are noted by many scholars (Altman, 1975; Sommer, 1969). For instance, Sommer

(1969) has distinguished these two concepts along the following lines:

1. Personal space is portable whereas territory is relatively stable and stationary.
2. The boundaries of personal space are invisible whereas the boundaries of territory do not.
3. Personal space has the person's body at the centre whereas territory does not.
4. Intrusion into personal space leads to discomfort and results in withdrawal tendency. In contrast, territorial encroachment leads to verbal threats and physical fights.

In this section attempt will be made to define and elaborate the salient aspects of personal space and thereafter the major ways of measuring personal space will be discussed. Furthermore, major research work done in the area of personal space will be discussed and theoretical framework will be presented. Finally, application of personal space research and its relevance to the growing field of environmental design will be illustrated.

Conceptualization of Personal Space

Katz (1937) used the term "Personal space" for the first time. However, it was Hall (1966) who was the first to coin the term "proximics" which is more or less synonymous to personal space. Various scholars have adopted the distance component of interpersonal relationship and behavioural aspect of physical space in their operational definitions of personal space (Hall, 1966., Sommer, 1969., Proshansky, Ittelson & Rivlin, 1970; Craik, 1970). Sommer (1969) defines personal space as "an area with invisible boundaries surrounding a person's body into which intruders may not come." In his famous book entitled *The hidden dimension,* Hall (1966) defines personal space as a small protective sphere or bubble that an organism maintains between itself and others. A number of scholars have severely criticized the concept of personal space (Patterson, 1975; Altman, 1975; Knowles, 1980). These scholars feel that the concept of personal space is misleading and inappropriate. Some of the criticisms are as follows:

1. The concept of personal space implies stability, while in fact it stretches and shrinks with environmental circumstances and relationship.
2. The personal space concept places heavy emphasis on distance and clearly excludes various aspects involved in social interaction like eye contact.
3. The concept of personal space exists only when interaction with others is involved. Thus, the concept of personal space is interpersonal rather than personal.

Despite criticism the concept of personal space is still very popular and well established in the literature of environmental psychology. Attempts are made to deal with some of the major criticism and to broaden the application of the personal space concept. For instance, the personal space is defined as an area with invisible boundaries surrounding a person's body and any intrusion by others is not welcomed. Gifford (1987) has emphasized the distance component of interpersonal relations in his operational definition of personal space. He suggested that personal space is both an indicator and integral part of growth, maintenance and decline of interpersonal relationships. Furthermore, some researches have suggested that the personal space bubble is not circular but elliptical. It is bigger in front and behind us than at the sides. Thus, it seems that people will generally tolerate others coming closer to us at the side rather than from front or behind (Cassidy, 1997).

In short, personal space is conceived as an envelope surrounding a person. The invasion of personal space by unwanted others leads to a general tendency of withdrawal and discomfort. Personal space is not fixed and stable but it changes considerably depending on context, interacting persons and circumstances. It is directly linked to the communication channel and thus it is dynamic and active concept of human spatial behaviour.

The Functions of Personal Space

The foremost guiding notion regarding the functions of personal space is related to appropriate interpersonal spacing. Invasion of personal space results from inappropriate interpersonal distance produces more than one adverse consequence like stress, anxiety and improper communication (Altman, 1975). Another important function of personal space is mechanism of communication and can be understood completely with other aspects of non-verbal communication such as orientation, touch, and eye contact (Cassidy, 1997). It was the influential book of Heider entitled *Wild animals in captivity* where identification of four distinct personal zones frequently used by animals is discussed. These zones were labelled as flight distance, critical distance, personal distance and social distance. The work of Heider stimulated Hall (1959) who developed an insightful framework of 'proximics'. The "proximics" approach emphasizes the ways of relating to the peoples and active involvement in the management of space to attain desired level of interpersonal interaction with others. The interpersonal distance between interacting parties is helpful in the understanding of the nature and level of relationships between them. In addition, personal space is to a very large extent a function of our relationship with other person involved and the society and culture to which we are accustomed. In essence, we use personal space to communicate our relationship with others, to protect our territory, and to generally regulate our interaction with others.

Hall (1966) proposes the use of four personal zones in social interaction between individuals. Each of which can be subdivided into near and far. A brief description of these zones is given below:

Intimate Distance

The interpersonal distance ranges from 0 to 18 inches. At this distance strong sensory inputs like smell, body heat, sound and a feeling of breath combine to create inescapable involvement between interacting parties. A typical behaviour at this distance is a very low level of voice. Hall distinguishes between near situations requiring body contact (love-making) and far distances which requires being very close but not in contact (whispering). This distinction is artificial because the contact depends on various aspects of social and physical setting.

Personal Distance

The distance between interacting parties range from 1.5 to 4 feet. Usually, this distance is used for most interactions between other people. Within this personal distance zone a 'close' phrase is used for good friends or couples and a 'far' phrase of interpersonal distance is used for friends and acquaintances involved in social interaction. Classmates usually interact in the 'far' phrase of personal space distance. The voice level between interacting parties is moderate and sensory inputs are at minimum.

Social distance

This distance zone range from 4 to 12 feet and is characterized by Hall as business distance. The transactions are formal between people who are not well-acquainted or business representatives and the level of voice is louder. The near distance would be used by those being introduced or for informal business transactions. In contrast, the far distance would be reserved for more formal business transactions or processes.

Public Distance

This distance zone extends beyond 12 feet. Usually, this distance is maintained between speakers and public. Examples of public distance are class lectures, public meeting of leaders. The level of voice is amplified.

The proximics framework of Hall is very well articulated and developed. The conception of four distance zones is appealing for it illustrates certain interpersonal communication possibilities. Furthermore, the framework of Hall has stimulated wealth of research but at the same time it is critically appraised. The four categories of personal space helps to illustrate the different functions of personal space are artificial in nature. In fact it is likely that personal space spans around vast continuum, determined by number of factors including relationship with other person, cultural and social norms and the physical setting. Some of the other critical remarks are as follows: First, the proximics framework is influenced by ethnological tradition. It places limited emphasis on the protective functions of personal space. In fact, depending on the situational context and type or level of relationship the personal space changes is not adequately pointed out by Aiello, (1987). However, the protective function of human spacing as noted by Hall has been adopted by many scholars working in the area of environmental stress and crowding (Baum & Paulus, 1987; Evans & Cohen, 1987). In addition, the work of Hall has stimulated a number of scholars to develop comprehensive models to explain the complex relationship between human spatial behaviour with non-verbal and verbal variables (Gifford, 1987; Altman, 1975).

As a cultural anthropologist Hall (1966) work carefully highlights the cultural variations in spatial distancing. He has noted that the physical distance did not make the four zones important. The distance defining different zones may differ across cultures. For example, Middle Eastern peoples tend to tolerate closer distances than people from Britain. In the next section an attempt will be made to describe major methods of measuring personal space. The other section will focus on some major studies conducted in the area of personal space followed by recent theoretical models and approaches that guides the human spatial behaviour.

Measurement Techniques of Personal Space

There are three general methods listed by Altman (1978) to measure personal space. These are: (1). Simulation or projective methods; (2) Laboratory or quasi-projective methods; and

(3) Field naturalistic or interactional methods. The most commonly used methods to measure personal space are based on simulation techniques, the stop distance and the naturalistic observation methods. A brief description of these methods is given below:

Simulation Methods

In a recent review on human spatial behaviour Aiello (1987) has reported that over forty percent of the personal space studies have used simulation methods. These methods require subjects to imagine some interactional situation and to reconstruct the distance between the self and another person from memory. Usually, the task of the subject is to arrange figures in some fashion that were shown to them for brief period of time by the instructor or to make judgement about the distance that separates the figures.

The simulation methodology was developed for the first time by Kuethe (1962) in their study on social schemes. Subjects were asked to place miniature figures representing the self and others on a board. Considerable consistency and organization in the individual arrangement of figures was reported. The methodology used by Kuethe stimulated a number of simulation techniques to measure personal space.

The comfortable interpersonal distance scale (CID) devised by Duke and Nowicki (1972) is a frequently used simulation method to measure personal space. In their paper and pencil measure the self is shown in the centre and eight lines radiate from the self in each direction. The subject is asked to mark on a prepared form the distance they would place themselves in relation to other person's approach, beyond which discomfort is likely.

Based on a review on personal space Hayduk (1983) points out some serious limitation of projective techniques and argues, "The projective techniques are inferior and unaccepted measure of personal space". In addition, an empirical study conducted by Love and Aiello (1980) reported that projective techniques lack construct validity. Several other studies have consistently noted that simulation methods do not measure the alpha personal space (external measurable objective distance between interacting individuals) and are found to show low correspondence between actual objective distances (Hayduk, 1983; Knowles, 1980). Gifford (1987) has suggested that these techniques are appropriate in the understanding of better personal space or the subjective experience of the distancing process.

Laboratory Methods

The most common laboratory method to measure personal space is the stop distance procedure. The participants are asked to approach at the point when feeling of discomfort is experienced. This procedure is repeated for different angles of approach. The interpersonal distance between participants provides the measure of personal space. The stop distance model is frequently used by scholars to measure personal space (Hayduk, 1983).

Laboratory studies tend to overcome some of the limitations of projective techniques. For instance, in stop distance method the subject is not required to imagine and reconstruct from memory the personal space distances. However, the limitation of using this method is that the subject's awareness of the distance process becomes very salient and fails to correspond with the distance chosen by them in their day-to-day lives.

Field Naturalistic Methods

These measures are less frequently used in personal space research because of various reasons. The most important reason seems to be related to the measurement of distance under natural conditions for they are subject to many unknown and uncontrollable influences like status of the partners, topics being discussed, levels of relationship. Another difficulty concerns with the appropriate estimation of the interpersonal distance between participants interacting in natural settings.

The naturalistic methods involve the unobtrusive

observation of participants engaged in actual interaction. One variant of naturalistic methods is to observe the participants personal space in the laboratory situation. The actual purpose of the investigation is generally concealed from the subjects. The experimenter asks the subjects to pull a chair from the stack of chairs kept in the corner of the room. The interpersonal distance between the experimenter and the subject seating can be measured which provides a measure of personal space. A number of related studies are discussed in the literature, where personal space is measured in the naturalistic settings in laboratory like situation (Gifford, 1987).

All the three methods of assessing personal space—simulation, laboratory and naturalistic methods have their advantages and limitations. The simulation and the laboratory methods are criticized for they tend to increase the salience of distance process and are low on external validity. The simulation method is best suited to assess beta personal space rather than alpha personal space because reconstruction from memory is required to scale down space between themselves and others. Usually, the laboratory methods is primarily used to study the live encounter between participants, but this method tends to ignore the role of various proprioceptive cues usually available in real life interactions. Finally, the naturalistic observation methods sound very promising for it intends to measure personal space in real life settings. However, assessment of personal space is technically very difficult in these settings. Depending on the nature and purpose of the study and keeping in mind the relative advantages over other methods, the best method should be adopted to estimate the personal space.

Studies on Personal Space

In this section an attempt is being made to highlight some of the major factors that influence the size and use of personal space. The major review on personal space is neatly divided into three categories: (1) Personal factors, (2) Contextual factors, and (3) Cultural factors. A brief discussion of these factors in determining personal space is made and adverse consequences of personal space violation will be examined.

Studies on Personal Factors

Various studies have demonstrated that there is an influence of individual's personal characteristics such as age, sex and personality on personal space. However, the precise role of a given personal factor is very difficult to ascertain. It can only be known when other influences like different situational context and other contributing factors that might influence the participants distancing process are either held constant or are partial out. The strategy adopted by various scholars to study single personal characteristics across a variety of situational context is to access the main effect of personal influences on personal space.

Of all the demographical characteristics age and sex are two variables that are found to be very potent in determining personal space. Hayduk (1983) has reported that personal space increases gradually with age. However, conflicting evidences exist that children prior to age 5 do show decrease in personal space (Burgess & McMurphy, 1982) while some researchers reported evidence of increase in interpersonal distances (Smetana Bridgeman & Bridgeman, 1978). Thus, the development process prior to age 5 is not very reliable (Altman, 1975).

Sufficient evidence regarding the gradual and systematic development of spatial norms from age 5 through middle adulthood is documented in the literature. The distance increases with age and normal personal space behaviour are developed by the time children reach the age of 12 (Evans & Howard, 1973; Aiello & Aiello, 1974). The study of Aiello and Cooper (1979) examined the interpersonal distances of same sex pairs of children. The results of this study reveal a consistent pattern of spatial norm development. The 7 years old child typically maintains a distance of less than one feet while 17 years old children typically interact at a distance of 2 feet. Supportive evidence is reported by several other scholars interested in the developmental pattern of human spatial behaviour (Thompson & Aiello, 1981; Jones & Aiello, 1973; Tennis & Dabbs, 1975).

Studies on gender differences suggest that men to men pairs have larger personal space areas followed by women to women pairs and men to women pairs (Goffman, 1963; Lott & Sommer, 1967; Gifford, 1987). These studies have reported that male/female combinations are generally in close proximity to each other than the pairs made up of the same sex (Kuthe, 1962). In one study on personal space Severy, Forsyth and Wagner (1979) examined the influence of sex, gender and race simultaneously. Sex was not found to be powerful determinant of personal space. It only influenced the personal space in interaction with age and race. Age was found to be a more powerful determinant of personal space than race and gender.

Personality correlates of various kinds like dominance, warmth, extroversion-introversion is found to associated with personal space (Wiggins, 1979). Sufficient evidence exists regarding the relationship between extroversion and personal space with extroverts showing tendencies of smaller personal space areas relative to introverts (Cook, 1970; Mehrabian & Diamond, 1971). The study conducted by Gifford (1982) reveals that extroverted and gregarious people require smaller personal space, while cold and quarrelsome people require a larger interpersonal distances. However, some scholars have noted that there is no consistent relationship between personality correlates and personal space (Hayduk, 1985).

Considerable evidence suggests personal space is affected by problem behaviour and social emotional disorders. It is generally found that violent criminals require larger interpersonal distances. For the psychiatric patients the maintenance of interpersonal distances depends on the specific symptoms and severity of the problem behaviour. Sommer (1959) is credited to have conducted the first empirical study of personal space for schizophrenics. Their results indicated that as compared to normal hospitalized and non-schizophrenics the schizophrenic's patients chose a more variable and random personal space zones. They sometimes solicit much greater seating distances and sometimes much smaller distance zones. Similar findings were reported by Horowitz (1968). In one study Horowitz (1968) examined the relationship between inappropriate selection of personal space zones and the degree of disturbance in schizophrenic patients. The researchers reported that the severely affected schizophrenic patients had very large personal space. However, as the degree of disturbance was reduced and the conditions of schizophrenic patients showed improvement, the personal space was reduced and became more comparable to that of non-patients. Consistent results obtained from a number of studies suggest that psychologically disturbed subjects having personality abnormalities need more personal space (Booraem & Flowers, 1972; Fischer, 1967).

In short, personal space may be determined in part from various personal characteristics of an individual like age, gender and personality. However, a full range of personal space tendencies may be known only when various factors related to personal influences are treated together in a single research tradition. In addition, the situational influences should also be taken into account to attain a more comprehensive picture regarding the selection of personal space zones.

Studies on Cultural Influences

Most of the observations made by Hall (1959) point out that space utilization varies across cultures are general and are directed towards supporting the existence of four spatial zones. The work of Hall provides a comprehensive descriptive analysis of various cultural factors and alludes to possible differences in the use of space across people of contact *versus* non-contact cultures. He proposed that people in 'contact cultures' (e.g. Southern Europeans, Latin Americans and Arabs) as compared to people in 'non-contact' cultures (e.g. North Americans and Northern Europeans) maintain closer interpersonal distances, high interpersonal involvement, more direct body orientation and large amount of visual gaze. A number of essential issue in intercultural differences with regard to the tendency to interpret the invasion of personal space as an indication of aggression were identified by Hall. Based on the review of studies Smith and Bond (1993) concluded that Hall's proposed cultural differences are well supported for those who know each other well but not when the interacting parties are strangers. It seems plausible to expect that different aspects of non-verbal communication coexist and thus firm conclusion regarding the relative importance of different aspects (e.g. close proximity, eye contact and touch) on discomfort seems difficult.

The descriptive account of Hall (1966) works on cultural influences on spatial behaviour has received empirical support from scholars (Watson & Graves, 1966; Mehrabian, 1969; Noesjirwan, 1978). However, contrary and even inconsistent results exists (Forston & Larson, 1968; Mazur, 1977). The Watson (1970) study provides a detailed account of male foreign students from a number of social contact and 'non-contact' cultures. Their results reveal marked differences in spatial behaviour. The students from 'contact' cultures not only displayed close proximity but they generally faced each other more directly, spent more time visually gazing into each other's eyes as compared to the students from 'non-contact' cultures. A review of Altman (1987) on the influence of culture provides a more thorough account of various issues with regard to differential use of space and cultural differences.

Most of the cultural studies have been carried out with those participants who visited foreign countries temporarily for some period. It is possible that these foreign students treated as potential subjects in the cultural studies may not be the true representatives of their native countries. They are presumably different from the majority of their country's population (e.g. higher socio-economic status, higher educational level). Thus, the results discussed on contact *versus* non-contact cultures may be interpreted with great caution. Despite methodological limitation researchers do provide some evidence that culture is a major modifier of personal space.

Studies on Situational Factors

The studies discussed in this section deals with the question as to how the participants personal space tendencies are influenced by the situational factors. A number of studies on situational effects on personal space are primarily focused around the social aspects rather than the characteristics of the physical setting. It is observed that where attraction between individuals is strong, the tone of interaction is friendly we are more willing to reduce our personal space requirements (King, 1966; Little, 1965). Similarly, where people dislike each other and the tone of interaction is unfriendly people enhance their personal space and sit further away from one another (Guardo & Meisels, 1971; O'Neal *et al.*, 1980). The review of Altman (1987) examines a number of studies dealing with interpersonal relationship and personal space. These studies found that when positive relationship is involved such as those with friends, small personal space zones are preferred relative to those who are only acquaintances. In addition, as compared to strangers people tend to use less space with acquaintances for which they feel positive about and are liked. Similar trend of results are obtained by other scholars (Aiello & Cooper, 1979; Burgess, 1983).

Larger personal space zones are required when participants are insulted or are placed in a competitive setting (O'Neal, Brunault, Carifo, Trautwine & Epstein, 1980; Ugwuegbu & Anusiem, 1982). Studies have also shown that participants maintain greater distances when other person is hostile, physically impaired or is a mental patient (Rumsey, Bull & Gahagan, 1982; Wolfgang & Wolfgang, 1971). Additional space is required for people who are stigmatized as insensitive and clearly wish to distance themselves from others (Worthinton, 1970). In addition, discrepancy on participant's status also leads to larger personal space zones (Gifford, 1982; Mehrabian, 1969).

The nature of the situation may also influence personal space. The study of Sommer (1969) provide evidence that individual tend to select closer seats and choose less direct orientation in cooperative situation. In contrast, if the situation is competitive in nature, interpersonal spacing is increased and angle of orientation is more direct. Similar results were obtained in other studies. For instance, people in cooperative situation generally select close interpersonal distances unless the competition requires interpersonal contact (Cook, 1970). A study was designed by Tedesco and Fromme (1974) where participants were required to interact with each other either in cooperative or competitive situations. The selection of interpersonal space was observed on the same participants who were asked to wait in another room. The result of this study revealed that participants who had been in cooperative situation chose a small distance relative to participants who previously encountered competitive situation. The effect of status and selection of interpersonal space between participants is also investigated by researchers. The greater the status difference between participants the larger the interpersonal space used. Burns (1964) demonstrated that in a filmed interaction in an office situation, someone standing further away when communicating a message was judged as subordinate. However, when the same individual stood closer when conveying a message the status was judged as equal.

Our choice of interpersonal distance is also influenced by our anticipation regarding the type of person with whom we expect to interact. We tend to select smaller distances when we anticipate meeting friendly and warm person (Kleck, 1969). Similarly, we are not likely to help persons who tend to invade our personal space (Konecni *et al.*, 1975).

Some studies have examined the influence of environmental factors on interpersonal spacing. In general people maintain more spacing in the corner of the room than in the centre of the room. Similar distances seem to be preferred in rooms with higher ceilings and in narrow room rather than in those rooms which have lower ceilings or are square in nature (Daves & Swaffer, 1971; Tennis & Dabbs, 1975).

Based on the results obtained from empirical investigations, it appears that nature of the situation seems to determine the selection of personal distance zones. It appears that people exhibit strong preference for more space in conditions when the overall supply of space is low. Stigmatized persons, discrepancies in the status of the participants and competitive nature of the situation are associated with larger interpersonal distances.

Consequences of Invasion of Personal Space

A number of studies are directed to examine the consequences of personal space violation. The focus of these studies is to examine the consequences of inappropriate physical space between individuals in a variety of naturalistic settings ranging from college libraries, shopping complexes, public parks, airports to a controlled laboratory conditions.

Evidence exist that invasion of personal space causes states of over arousal that are experienced as stressful (Stokols, 1976; Evans, 1978; Gifford, 1987). Middlemist, Knowles and Matter (1976) conducted a study in three urinal lavatories. Unobtrusive observation was made by a hidden camera to measure how long it took to begin urination and duration of urination. These observations were made in three conditions. In the first condition the confederate stood close to the urinal point. In the second condition the confederate used the adjacent urinal while in the last condition no confederate was present. The results of this study revealed that closer invasion increased delay as well as amount of urination. Similarly, the study conducted by Reid and Novak (1975) on students and faculty revealed that subjects preferred the urinal farthest from the confederate. In one study conducted on male college students Barefoot, Hoople and McClay (1972) observed that subjects were less likely to drink and the duration of drinking from the water fountain was less when confederate was seated at one feet from the fountain as compared to five or ten feet of distance. Thus, it seems that invasion of personal space resulting in close proximity created arousal and tension in the subjects. The invasion of personal space is likely to increase heart rate, elevates the GSR and often generates a tendency of flight or avoidance.

Inappropriate spacing typically associated with elevated physiological arousal and discomfort (McBride, King & James, 1965; Middlemist, Knowles & Matter, 1976) often lead to various forms of compensatory reactions like turning away, avoiding eye contact and withdrawal behaviour (Felipe & Sommer, 1966; Patterson, Mullens & Romano, 1971). For instance, Patterson *et al.* (1971) measured latency of flight and compensatory behaviour of college students in library. Experimenter invaded the personal space of subjects by selecting their seat one, two or three seats away. The results of this study indicated that personal space invasion did produce flight behaviour. However, compensatory reactions were found to increase with immediacy of reaction. In one study latency of fleeing from college library was measured. The confederate either took the permission to invade or did not take permission from the subjects. It was found that females fled more quickly than males, especially when they asked permission from the invader (Aiello, 1987).

The study of Konecni, Libuser, Morton and Ebbesen (1975) examined the speed of pedestrians crossing the street after being invaded and the desire to help the invader. The findings of this study reveal that closer invasion resulted in faster walking than farther invasion. In addition, moving pedestrians whose space was invaded helped less when the dropped object was an important one. The results of Konecni *et al.* (1975) cannot be treated as inconclusive for the relationship between interpersonal distance and helping behaviour are not simple and may be affected by various factors like different attributions by the invaded subjects (Baron & Bell, 1976).

There are various situations in which personal space violations are perceived as less aversive and even accepted. For instance, Fry and Willis (1971) studied the reactions of male and female adults standing in line for a movie. The children of five, eight and ten years of age were instructed to move as close as possible without touching and conversing with adults in line. An intrusion of ten years old elicited negative reactions (e.g. moving away, leaning) for adults. Other studies have found the invasion as less aversive in various situations like if invading confederate is well dressed or has a very good reason for the invasion and has no choice but to invade. The uniqueness of such situations is accepted in many cases (Smith & Knowles, 1979; Murphy-Berman & Berman, 1978).

Theoretical Perspectives on Personal Space

Beginning with Hall's (1966) analysis of proximics framework various theoretical ideas have enriched the area of personal space. A brief analysis of major theoretical perspectives is presented below:

Intimacy Equilibrium Model

Largely adapted from the stimulus-response approach to learning (Miller, 1944) the intimacy equilibrium framework of personal space was developed by Argyle and Cook (1976). This model is an insightful extension of affiliative-conflict theory propounded by Arglye and Dean (1965). This model stresses on the existence of two conflicting behavioural tendencies—approach and avoidance that simultaneously attracts and repels participants in every interpersonal encounter. Furthermore, the prominent focus of this theory is on the development of equilibrium or affiliative balance which is a joint function of interpersonal distance and various non-verbal behaviours like eye contact, smiling, intimacy of the discussion topic and so forth. Individuals strive to maintain a state of equilibrium between approach and avoidance tendencies. Any change in immediacy behaviour creates a situation of disequilibrium often leading to discomfort. To restore desired level of intimacy and to attain equilibrium interacting participants try to change one immediacy behaviour or another. For instance, the resulting disequilibrium in the interpersonal distance can be successfully compensated for by shifting the topic of conversation, reducing eye contact, shifting body position and so forth.

The theory has received empirical support from a number of studies (Aiello & Thompson, 1980., Schneider & Hanswick, 1977). The multifaceted model of Argyle and Dean (1965) talks vaguely about the role of several immediacy behaviours, which determine the individual's responses to disequilibrium. In addition to this the theory falls substantially short in explaining the changing interpersonal relationships between interacting participants.

Extension of Intimacy Equilibrium Model

To overcome the limitations of affilative-conflict theory, several investigators have attempted to explain the process regarding changes in relationships. For instance, in social penetration theory, a dialectical approach is adopted which posits that compensatory and reciprocal behaviours occur as participants gradually adjust to changes in their attempt to attain a desired quota of interpersonal interaction (Sundstrom & Altman, 1976; Altman, Taylor & Wheeler, 1971). The conflict model of Kaplan (1977) treats attraction as a mediator and explains the individual's likelihood of reciprocity or compensation. If the initial interaction between participants is positive, the steeper shift in equilibrium towards the other will be seen and *vice versa.* Thus, this model explains not only why equilibrium levels change but also how in certain situation one person reciprocates other person approach behaviour instead of compensating for it. One shortcoming of this model is its failure to explain curvilinear relationship between interpersonal distance and various types of immediacy behaviour. In addition, only post-hoc explanation is possible and the prediction of behaviour from this model cannot be precisely made because the measurement of approach and avoidance gradients is difficult.

The curvilinear 'comfort' model (Aiello & Thompson, 1980) represents an extension of intimacy equilibrium model (Argle & Dean, 1985), which posits limits of compensation. The model suggests two important points. First, when only minor variation occurs in immediacy behaviours the compensatory process might not be visible. Secondly, beyond the prescribed limits of compensatory range there exist a critical zone of discomfort and reestablishment of a desired level of intimacy becomes very difficult. The study carried out by Thompson & Aiello (1981) reported evidence regarding the levels of comfort as a function of different interpersonal distances. Maximum comfort was reported at moderate distance while decline in comfort was found at shorter or longer distances. Evidence exists regarding different equilibrium points of physical proximity across male and female subjects. At inappropriate close distance males reflect more discomfort while females report more discomfort when distance becomes inappropriately far.

Arousal-Cognition Model

The arousal model of Patterson (1975) attempts to overcome one major shortcoming of equilibrium model (Arglye & Dean,

1965), which is related to inadequate focus on the aspects of non-verbal reciprocation and overemphasized on compensatory behaviours to maintain the disturbed intimacy level of participants. The model of Patterson (1975) postulates that any interpersonal encounter between participants involves changes in arousal, which has the cognitive component. Individual tend to label the changes in arousal as positively or negatively. In situations that induce arousal shifts accompanied by positive affect the behavioural tendencies of reciprocation is intensified. In contrast, negatively labelled arousal shifts lead to compensatory behaviours like withdrawal tendencies in participants.

While the arousal-cognition model of Patterson (1975) has received some empirical support (Foot, Chapman & Smith, 1977; Smith & Knowles, 1979), a number of scholars have criticized the model on several grounds. For instance, some scholars have stated that the cognitive-arousal model fails to provide a convincing account regarding the appropriate causes of positive or negative evaluations and that Patterson overemphasizes arousal in his model (Hayduk, 1983). In addition, the self-focused cognitive tendencies suggested by the model often misattribute the source of discomfort in the participants.

Expectancy and Discrepancy Models

These models are probably the most developed of all the models described before for they deal with the full range of interpersonal interactions between participants. Two major expectancy models—one proposed by Patterson (1975) and other by Cappella and Greene (1982) are probably the most comprehensively developed of all the expectancy models. These models suggest that the expectancy level is the combination of various situational factors, norms and personality characteristics of the interacting parties. Both the theories are interpersonal in nature and emphasize the process of non-verbal exchange. These theories effectively explain positive relationship obtained between interacting parties and non-verbal behaviour (reciprocation) but they also explain negative relationships (compensation) and non-linear relationships. The stable and unstable exchanges are directly linked with the degree of discrepancy between participants, expected interaction and actual interactions.

The above mentioned theoretical model represents the best potential for explaining the complexity of interpersonal interactions. However, various criticisms regarding the objective assessment of the internal characteristics exist. Only some of the external situational factors can be measured and hence the prediction of behaviour is not accurate besides these models are not comprehensively tested.

Application of Personal Space Research

The research on personal space has provided sufficient evidence that individuals are very sensitive to their spatial needs. Certain spatial information can be worked out into the internal designing of the built environment. However, largely depending on what behavioural pattern is valued in a particular setting, different provisions regarding the type of furniture arrangement is required so as to facilitate the attainment of user's need effectively. Osmond (1957) differentiated the environmental settings in two distinct categories—the sociopetal and the sociofugal settings. The sociopetal arrangement (e.g. chairs arranged in circles or oval) orient individual towards one another and facilitates interaction. In contrast, the sociofugal seating arrangements tend to inhibit interpersonal interaction among individuals. The airport-waiting lounge is a typical example of a sociofugal setting where chairs are bolted together in rows. The waiting passengers are forced to orient themselves outward and away from one another, which discourages the probability of interaction with other fellow passengers. Sufficient empirical evidence exists that waiting in dental clinics or waiting for a delayed flight becomes more uncomfortable largely because of the sociofugal nature of the internal seating arrangement (Sommer, 1967).

A growing body of work conducted by scholars in the reading room of college libraries also seems to provide evidence regarding the tremendous impact of internal designing needs and resulting affective states. Some of the observations noted by Aiello (1987) can be used in redesigning the spatial arrangements of the library room's set-ups. Some suggestions for renovating the structure of the library rooms consistently with the user's need are as follows:

1. The students usually prefer library seats, which allow back to back seating arrangements relative to side-by-side arrangements.
2. If the tables are occupied, students tend to select those seats, which are located further away from the occupants. The subjects generally prefer seats where tables were occupied.
3. When the room is over sixty per cent full the users will avoid sitting in library and they may move to their dorms or cafeterias.

Granted the fact that interaction has become a vital part of our everyday life, we need to be aware of the potential positive and negative effects of the use of interpersonal distance. The idea generated from personal space research does provide some useful leads to architectural designers. The knowledge regarding the spatial needs of different group of users is essentially to enhance satisfaction and facilitates desirable behaviours of users. Many scholars note this fact across different settings like college libraries, hospitals, airports, restaurants, counselling centres. Thus, it seems very clear that use of interpersonal distance is an important aspect of the regulation of interaction and its effectiveness.

Territorial Behaviour

In the beginning of this chapter a difference was made between personal space and territorial behaviour. The concept of territory was originally used to understand the animal behaviour. The acquisition, demarcation and defence of territory is essential for the survival of animals in terms of security, provision and sometimes accumulation of food and mating. One can observe territorial behaviour in domestic animals. For instance, a cat often marks the boundary of its territory by urinating around the borders and attempts to defend against any intrusion. Wars are fought between nations to defend the territories. Often people build the boundaries of their houses, erect fences and markers to communicate to others their territorial rights and show resentment and irritation, in case of encroachment. Often students leave their bags or belongings on tables and chairs to mark their territory which often is respected by others. According to Fischer, Bell and Baum (1984) "human territoriality can be viewed as a set of behaviour and cognitions an individual or group exhibits based on perceived ownership of physical space." Territorial behaviour serves important individual needs and motives which include controlling, personalizing and defending an area.

Types of Territory

Altman and Chemers (1980) have given one of the most commonly used categorization for defining and measuring the territorial behaviour. They labelled the three types of territory as primary, secondary and public that tends to differ along the dimensions like the duration of occupancy and psychological centrality. The ***primary territory*** refers to place that is relatively permanently owned either by an individual or an interdependent group. Primary territories are typically occupied for long periods and are central to the lives of the owners. People usually give lot of psychological importance to their primary territory. Any intrusion in this territory provokes high arousal and the defenses used by the individual occupying the territory are very strong. The owner has full control and intrusion is considered as a serious matter. Disregarding the size some of the examples of the primary territory are one's home, bedroom, or nation. The ***secondary***

territories are less central to a person and are generally owned by the individual on a temporary basis (e.g. classroom, club, bar). These territories are more accessible to a greater range of users, but regular occupants exert some control over who may enter the territory and what range of behaviour may take place. The occupants are perceived as one potential and qualified user. The secondary territory may be personalized to some extent during the period of legitimate occupancy giving the occupant some legitimate power. However, the time spend by the occupants in the secondary territories is usually limited relative to the time spend in the primary territory. The limits of occupancy are determined not solely by the users, but by the collective owners of the secondary territories. The difference between the primary and secondary territory is regarding how much perceived importance an individual gives to the place. Finally, the ***public territories*** are generally accessible to everyone and are not owned by any one person (e.g. sea beach). An individual is considered as one of the potential user and there is little likelihood of defence in case of intrusion.

Because the type of territory is not defined in clear-cut physical terms, conflicts are sometimes seen regarding the ownership of a space. For instance, a frequent user of the library may come to define a particular seat as a secondary territory. However, an infrequent user may occupy the same seat and may face resentment from the daily user. The residents residing in high-rise complex may define the garden, swings and pool facilities as their secondary territory and may act defensively to strangers who use the garden as a public territory.

Territorial behaviours are common both in human and animal population. Proponents adhering to the sociobiological perspective argue that territorial behaviour is inherited and instinctive. Both animals and humans have a drive to own and defend their territory (Ardey, 1966; Lorenz, 1966). Others who believe in the perspective of cognitive psychology argue that territory serves an organizing function and is learned. Territorial behaviour is a function of our experience in the world and is very much dependent on social and cultural factors. Finally, some researchers consider territorial behaviour is the result of interaction of instinct and learning (Alland, 1972; Klopfer, 1968). However, the role of both instinctive and learning processes is not clearly specified (Altman & Chemers, 1980). It is possible that human beings are predisposed towards territorial behaviour through instinct, but their learning determines the intensity and form of territorial action. Another perspective is to assume that instinct guides some type of elementary territorial behaviour. The complex behavioural patterns are guided by the processes of learning is noted by scholars (Aiello, 1987).

Functions of Territorial Behavior

One of the major functions served by the territorial behaviour in humans is organizing (Taylor & Stough 1978; Edney, 1975). A comprehensive outline regarding organizing functions of human territories in everyday settings given by Taylor (1978) is illustrated below:

For people in...	*Organizing Functions of Territory*
Public Places (e.g. a library, the beach)	Organizes space, provides an interpersonal distancing mechanism.
Primary territories (e.g. bedroom)	Organizes space by providing a place which promotes solitude, allows intimacy and expresses personal identity.
Small face-to-face groups (e.g. the family)	Clarify the social ecology of the group and facilitates group functioning may provide home court advantage.
Neighbourhood and Communities	Promotes an 'in-group' who belong and can be trusted, differentiates them from 'out-group'. In some urban areas, territorial control makes a space safe to use.

As can be inferred from the framework of Taylor and Stough (1978) territories facilitates the mapping of different types of behaviour to help us plan and order our daily lives. Furthermore, territorial behaviour provides us with a sense of identity. We have heard people of high status stating their feeling with great deal of proud that they are in possession of a given territory. It provides us with a feeling of identity. Territorial behaviour is helpful in promoting predictability, order and stability in life. We can anticipate whom we are going to encounter in a particular territory and what will be the likely status of people who are visiting the given territory.

With regard to territorial defence a number of dissimilarities are reported between humans and animals. Disregarding the international conflict or territorial disputes over a piece of land between opposing parties, humans rarely resort in actively defending their territories. One reason for the lower degree of territorial defense in human relative to animals is that humans are rational and logical and thus very often they respect the territorial rights of others. Furthermore, humans very frequently invite and entertain others on their turf with pleasure, while the obverse is true for animals. In extreme cases the territorial fights between nations is often at a group level (animals territorial defences are more individualistic) where sophisticated weapons are used from long distances without entering the territory.

Research Studies on Territoriality

Factors Influencing Territorial Behaviour

A number of studies have examined the effect of group size, culture and gender on territorial occupancy, intrusion, and defence. With the increase in occupancy time in beach territories, the same sex groups claim larger spaces while mixed-sex groups use more markers. It seems that the mixed-sex groups display markers to keep their territorial boundaries intact. In contrast, the same-sex groups keep the boundaries unmarked so as to socialize outside the boundaries. Evidence exist that with the increase in group size, the average space per person decreases (Edney & Jordan-Edney, 1974), although this may be true for groups of friends, but not for strangers (Edney & Grundmann, 1979). Thus, territorial behaviour is influenced by personal and situational factors. It appears that males prefer larger territories in beaches and dormitories relative to females (Gifford, 1987).

Investigators have examined the ways in which people from different nationalities compete for beach space particularly in holidays. The result of the study conducted by Smith (1981) on Germans, French and Americans reveal that males claimed more space relative to females regardless of their nationality, and groups tended to occupy less space per person than couples or individual occupant. Furthermore, some differences were also reported which suggested that Germans engaged in more territorial marking by erecting boundaries and claimed larger spaces relative to their French and American counterparts. Finally, similarities in shapes were observed. However, the marking made by individuals were elliptical and groups marking territories where in circular shapes. With regard to sex differences the study conducted by Shafferand Sadowski (1975) reveal that male markers in a bar are intruded on fewer trials and, when intruded on, protect the territory for a longer period of time relative to females. Furthermore, in no case a lone intruder invaded the male territory. In one study Haber (1980) instructed confederates to intrude the classroom desk territories of male and female students. The confederates were not able to complete the planned intrusion when the occupant of the territory was male. In contrast, the confederates were able to complete twice the number of intrusion. It seems that confederates were reluctant to invade the male territories. The males were more likely to say unfriendly things to the invaders when cafeteria tables were intruded (Calsyn, 1976). Other scholars found that, when invaded, both males and females were equally likely to withdraw their library seats (McAndrew, Ryckman, Horr & Solomon, 1978). Thus, territorial behaviour does show some variation between cultures and gender effects.

Perhaps in many situations like protecting seats at a game or movie theatre by markers are less effective. In such public places the occupant while temporarily vacating the seat request the neighbours to save seats. Sommer and Becker (1969) observed that neighbours were more likely to defend an adjacent space when the duration of user's absence was short (e.g. 15 min *vs.* 60 min). Likelihood of defence was unrelated to the user's prior duration of occupancy (5 min *vs.* 20 min), or degree of personalized contact with the neighbour (no contact *vs.* asking for time *vs.* chatting information, all for 5 min). However, when neighbours were explicitly requested to defend the territory, as many as 71 per cent defended the territory by putting their own markers on the claimed seats (Hoppe *et al.*, 1972).

Territorial Demarcation, Personalization, and Behaviour

Territoriality is expressed through marking and often involves personalization. According to Cassidy (1997) fences, boundaries, and other indications that territory belongs to someone, without identifying who the person or group is, are markers. Markers with the identity of the person such as name plates or which distinguish the territory in unique way (e.g. house name or number) reflects attempt to personalize the territory. Public territories are self-consciously marked to reserve space. People often mark territories in public domain (e.g. seat in the library or space in the beach) so as to prevent intrusion and distress. Often when the territory in the public domain is invaded, people move their occupancy to other places. However, the invasion generates certain physiological arousal and anger (Cassidy, 1997). In contrast, it is less likely that the primary territory (e.g. home, garden) is invaded.

The effectiveness of public territory markers is dependent on many factors including duration of absence of the owner, the availability of alternative space, and the desirability of the territory (Brown & Altman, 1983). When sufficient space is available the territorial markers are effective and intrusion is prevented. However, when the alternative space decreases with increase in the population the territorial markers lose its significance and effectiveness (Sommer & Backer, 1969). Similarly the effectiveness of a marker (e.g. school newspaper on snack bar table), is reduced, if it is mistakenly perceived by the potential intruder as Litter (Arenson, 1977). However, markers such as college notebooks are not disregarded completely.

Some studies have reported that a half-empty beer mug is a more effective deterrent to intrusion of a table at a bar than a more personal marker such as sport jacket (Hoppe, Greene & Kenny, 1972). In a review Brown and Altman (1983) reasoned that prospective customers realize that patrons may forget their jacket when leaving the bar but they rarely leave half empty bear mugs in the bar because the prospective customers assume that the bartender expects the patron to return. Thus, personal markers provide more protection from intrusion than impersonal markers.

Territoriality Behaviour and Aggression

Depending upon a number of situational factors, territory may serve either as instigator of aggression or as a stabilizer of aggression. The status of a particular territory affects the relationship between territoriality and aggression. When territory is unestablished or disputed, aggression is more likely. The study of Lev and Cybriwsky (1974) reveal that street gangs engaged in more inter-gang violence when territorial boundaries were ambiguous than when they were well-established. Furthermore, evidence exist that well-demarked territorial boundaries lend stability and reduces the probability of violence and aggression. Other studies reported that after the introduction of identifiable territories a drop in the level of aggression in the group of retarded boys was seen. Thus, it seems that well-established territories are less likely to be intruded and thus the chances of aggression are less (Aiello, 1987).

Architectural Features and Territorial Behaviour

In public spaces certain architectural features appear to encourage the development of territorial claims. The study of Sommer (1967) reports that students are more likely to make territorial claims for library seats that are close to the wall, facing away from the distractions of the main entrance, and towards the rear of the room. Passerby is more likely to use water fountain if it is shielded from the spatial proximity of others by a physical barrier (Aiello, 1987).

It is commonly observed that much of the criminal activities are centred in areas around public spaces. The offenders are more likely to position themselves in street corners or pathways in open spaces and try to avoid front garden spaces. The concept of defensive space focuses on various aspects of public space organization that helps in promoting sense of ownership to its residents (Newman, 1972). Newman (1972) conducted a study on two housing projects in New York, occupied by equal number of residents. The Brownsville project was organized in smaller blocks having residential units for five to six families and was built around courtyard .The Van Dyke project was high-rise with residential blocks, separated by large parks. The design features of the Brownsville project provided defensive space to the residents. The design features facilitated sense of community and social interaction between neighbours in Brownsville housing, which promoted security and reduced the criminal mishappenings. In contrast, the crime rate was 50 per cent higher in Van Dyke housing relative to Brownsville because the probability of social interaction was inhibited by the residential design. Thus, it seems that the strength of territorial ownership and defence seems to vary according to the characteristics of the setting, the features of architectural design, gender, cultural background and group size.

Summary

This present volume has reviewed research on three major issues to the domain of human spatial behaviour. While two interrelated issues—personal space and territorial behaviour are discussed in this chapter, the research on crowding is reviewed in chapter five of this volume. The concepts of personal space and territory are interrelated. Thus, conceptual and methodological differences between these two concepts are discussed briefly.

The chapter provides an overview of various conceptualizations and functions of personal space. Personal space is invisible, mobile and body centred, regulating how closely we interact. In this continuation Hall's Proxemics framework is discussed which is based on the use of four personal zones in social interaction between people. The two major functions of personal space are protection and communication. The chapter then examines the issue dealing with three major measurement techniques of personal space. The applications of simulation, stop distance, laboratory and field naturalistic techniques are discussed.

The amount of personal space we maintain with others varies as a function of personal, situational and cultural factors. A brief review of these factors in the research on personal space is discussed. In the next section the consequences of personal space violation in different contextual settings is evaluated. An analysis of the major theoretical frameworks in the area of personal space is presented. While the framework presented by Hall (1966) is presented in the preceding section of this chapter, other theoretical perspectives related to Intimacy Equilibrium Model and its extension and expectancy discrepancy model is presented. Finally, application of personal space research is presented.

Territory is visible and stationary. It regulates our interaction pattern with others. The characteristics of primary, secondary and public territories and its organizing functions are discussed. Depending on the situation territorial invasion often lead to different degrees of aggression. The chapter further discusses how territoriality is expressed through marking and often involves personalization. Finally, the chapter ends with a brief discussion on architectural features and territorial behaviour.

REFERENCES

Aiello, J.R. (1987). Human spatial behavior. In D. Stokols and I. Altman (Eds.), *Handbook of Environment..l Psychology*, John Wiley & Sons, New York.

Aiello, J.R., & Aiello, T.D. (1974). The development of personal space: Proxemic behavior of children 6 through 16. *Human Ecology*, 2, 177-189.

Aiello, J.R., & Cooper, R.E. (1979). *Personal Space and Social Affect: A Developmental Study*. Paper presented at the Meeting of the Society for Research in Child Development, San Francisco.

Aiello, J.R., & Thompson, D.E. (1980). When compensation fails: Mediating effects of sex and locus of control at extended interaction distances. *Basic and Applied Social Psychology*, 1, 65-82.

Alland, A. (1972). *The Human Imperative*. New York: Columbia University.

Altman, I. (1978). Crowding: Historical and contemporary trends in crowding research. In A. Baum & Y.M. Epstein (Eds.), *Human Response to Crowding*. Hillsdale, N.J.: Erlbaum.

Altman, I. (1975). *The Environment and Social Behavior*. Monterey, Calif: Brooks/Cole.

Altman, I., & Chemers, M. (1980). *Culture and Environment*. Monterey, Calif: Brooks/Cole.

Altman, I., Taylor, D. A., & Wheeler, L. (1971). Ecological aspects of group behavior in social isolation. *Journal of Applied Social Psychology*, 1, 76-100.

Ardey, P. (1966). *The Territorial Imperative*. New York: Atheneum.

Arenson, D.D. (1977). Reactions to invasions of market seats at a racetrack. *Social Behavior and Personality*, 5, 225-228.

Argyle, M., & Cook, M. (1976). *Gaze and Mutual Gaze*. Cambridge, England: Cambridge University Press.

Argyle, M., & Dean, J. (1965). Eye contact, distance and affiliation. *Sociometry*, 28, 289-304.

Barefoot, J.C., Hoople, H., & McClay, D. (1972). Avoidance of an act which would violate personal space. *Psychonomic Science*, 28, 205-206.

Baron, R.A., & Bell, P.A. (1976). Physical distance and helping: Some unexpected benefits of "crowding in" others. *Journal of Applied Social Psychology*, 6, 95-104.

Baum, P., & Paulus, P.B. (1987). Crowding. In D. Stokols & I. Altman (Eds.), *Handbook of Environmental Psychology* (pp. 533-570). New York: Wiley.

Booraem, C.D., & Flowers, J.V. (1972). Reduction of anxiety and personal space as a function of assertion training in severely disturbed neuropsychiatric inpatients. *Psychological Reports*, 30, 923-929.

Brown, B.B., & Altman, I. (1983). Territoriality, street form, and residential burglary: An environmental analysis. *Journal of Environmental Psychology*, 3, 203-220.

Burgess, J.W. (1983). Interpersonal spacing behavior between surrounding nearest neighbors reflects both familiarity and environmental density, *Ethology and Sociobiology*, 4, 11-17.

Burgess, J.W., & McMurphy, D. (1982). The development of proxemics spacing behavior: Children's distances to surrounding playmates and adults change between 6 months and 15 years of age. *Developmental Psychobiology*, 15, 557-567.

Burns, TY. (1964). Non-verbal communication. *Discovery*, 31-35.

Calhoun, J.B. (1962). Population density and social pathology. *Scientific American*, 206, 157-163.

Calsyn, R. (1976). Group responses to territorial intrusion. *Journal of Social Psychology*, 40, 51-58.

Cappella, J.N., & Greene, J.O. (1982). A discrepancy arousal explanation of mutual Influence in expressive behavior in adults and infant-adult interaction. *Communication Monographs*, 49, 89-114

Cassidy, T. (1997). *Environmental Psychology: Behavior and Experience in Context*. Sussex: Psychology Press.

Cook, M. (1970). Experimentation in orientation and proxemics. *Human Relations*, 23, 61-76.

Craik, K.H. (1970).The environmental dispositions of environmental decision makers. *Annals*, 389, 80-94.

Daves, W.F., & Swaffer, P.W. (1971). Effect of room size on critical interpersonal distance. *Perceptual and Motor Skills*, 6, 926.

Duke, M.P., & Nowicki, S. (1972). A new measure of social learning model for interpersonal distance. *Journal of Experimental Research in Personality*, 6, 119-132.

Edney, J.J., Walker, C.A., & Jordan, N.L. (1976). Is there reactance in personal space? *Journal of Social Psychology*, 100, 207-217.

Edney, J.J., & Grundmann, M.J. (1979). Friendship, group size, and boundary size: Small group spaces. *Small Group Behavior*, 10, 124-135.

Edney, J.J., & Jorden-Edney, N.L. (1974). Territorial spacing on the beach. *Sociometry*, 37, 92-104.

Evans, G., & Howard, R.B. (1973). Personal space. *Psychological Bulletin*, 80, 334-344.

Evans, G. (1978). Human spatial behavior: The arousal model. In A. Baum & Y.

Epstein (Eds.), *Human Response to Crowding* (pp. 283-303). Hillsdale, NJ: Erlbaum.

Evans, G., & Cohen, S. (1987). Environmental stress. In D. Stokols & I. Altman (Eds.), *Handbook of Environmental Psychology*, New York: Wiley.

Felipe, N., & Sommer, R. (1966). Invasion of personal space. *Social Problems*, 14, 206-214.

Fisher, R.L. (1967). Social schema of normal and disturbed school children. *Journal of Educational Psychology*, 58, 88-92.

Foot, H.C., Chapman, A.J., & Smith, J. R. (1977). Friendship and social responsiveness in boys and girls. *Journal of Personality and Social Psychology,* 35, 401-411.

Forston, R.F., & Larson, C.U. (1968). The dynamics of space: An experimental study in proxemic behavior among Latin American and North Americans, *Journal of Communication,* 18, 109-116.

Fry, A.M., & Willis, F.N. (1971). Invasion of personal space as a function of age of the invader. *Psychological Record,* 2, 385-389.

Gifford, R. (1982). Projected interpersonal distance and interpersonal choices: Personality, sex, and social situation. *Social Psychology Quarterly,* 45, 145-152.

Gifford, R. (1987). *Environmental Psychology: Principles and Practice.* Boston: Allen & Bacon.

Goffman, E. (1963). *Behavior in Public Places.* New York: Free Press.

Guardo, C.J., & Meisels, M. (1971). Child-parent spatial patterns under praise and Reproof. *Developmental Psychology,* 5, 365.

Haber, G.M. (1980). Territorial invasion in the classroom: Invadee response. *Environment and Behavior,* 12, 17-31.

Hall, C.T. (1959. *The Silent Language.* New York: Doubleday.

Hall, C.T. (1966).*The Hidden Dimension.* New York: Doubleday.

Hayduk, L.A. (1983). Personal space: Where we now stand. *Psychological Bulletin,* 94, 293-335.

Hayduk, L.A. (1985). Personal space: The conceptual and measurement implications of structural equation models. *Canadian Journal of Behavioral Science,* 17, 140-149.

Hoppe, R. A., Greene, M. S., & Kenny, J. W. (1972). Territorial markers: Additional findings. *Journal of Social Psychology,* 88, 305-306.

Horowitz, M.J. (1968). Spatial behavior and psychopathology. *Journal of Nervous and Mental Disease,* 146, 24-35.

Jones, S.E., & Aiello, J.R. (1973). Proximic behavior of black and white first, third, and fifth-grade children. *Journal of Personality and Social Psychology,* 25, 21-27.

Kaplan, K.J. (1977). Structure and process in interpersonal "distancing". *Environmental Psychology and Nonverbal Behavior,* 1, 104-121.

Katz, D. (1937). *Animals and Men.* New York: Longmans, Green.

King, M.G. (1966). Interpersonal relations in preschool children and average approach distance. *Journal of Genetic Psychology,* 109, 109-116.

Kleck, R. (1969). Physical stigma and task oriented interaction. *Human Relations,* 22, 53-60.

Klopfer, P.H. (1968). From Ardrey to altruism: A discourse on the biological basis of human behavior. *Behavioral Science,* 13, 399-401.

Knowles, E.S. (1980). An affiliative conflict theory of personal and group spatial behavior. In P.B. Paulus (Ed.), *Psychology of Group Influence.* Hillsdale, NJ: Erlbaum.

Konecni, V.J., Libuser, L., Motron, H., & Ebbesen, E.B. (1975). Effects of a violation of personal space on escape and helping responses. *Journal of Experimental Social Psychology*, 11, 288-299.

Kuethe, J.L. (1962). Social schemas. *Journal of Abnormal and Social Psychology*, 64, 31-38.

Ley, D., & Cybriwsky, R. (1974). The spatial ecology of stripped cars. *Environment and Behavior*, 6, 53-68.

Little, K.B. (1965). Personal space. *Journal of Experimental Social Psychology*, 1, 237-247.

Lorenz, K. (1966). *On Aggression*. New York: Harcourt Brace Jovanovich.

Lott, B.S., & Sommer, R. (1967). Seating arrangements and status. *Journal of Personality and Social Psychology*, 7, 90-95.

Love, K.D., & Aiello, J.R. (1980). Using projective techniques to measure interaction distance: A methodological note. *Personality and Social Psychology Bulletin*, 6, 102-104.

Mazur, A. (1977). Interpersonal spacing on public benches in "contact" vs. "non-contact" cultures. *Journal of Social Psychology*, 101, 53-58.

McAndrew, E.T., Ryckman, R.M., Horr, W., & Solomon, R. (1978). The effects of invader placement of spatial markers on territorial behavior in a college population. *Journal of Social Psychology*, 104, 149-150.

McBride, G., King, M.G., & James, J.W. (1965). Social proximity effects on galvanic skin response in adult humans. *Journal of Psychology*, 61, 153-157.

Mehrabian, A. (1969). Significance of posture and position in the communication of attitude and status relationships. *Psychological Bulletin*, 71, 359-372.

Mehrabian, A., & Diamond, S.G. (1971). Effects of furniture arrangement, props, and personality on social interaction. *Journal of Personality and Social Psychology*, 20, 18-30.

Middlemist, R.D., Knowles, E.S., & Matter, C.F. (1976). Personal space invasion in the lavatory: Suggestive evidence for arousal. *Journal of Personality and Social Psychology*, 33, 541-546.

Miller, N. (1944). Experimental studies in conflict. J. McV. Hunt (Ed.), *Personality and the Behavior Disorders* (Vol. 1). New York: Ronald.

Murphy-Berman, V., & Berman, J. (1978). The importance of choice and sex in invasions of interpersonal space. *Personality and Social Psychology Bulletin*, 4, 424-428.

Newman, O. (1972). *Defensible Space*. New York: Macmillan.

Noesjirwan, J. (1978). A laboratory study of proxemic patterns of Indonesians and Australians. *British Journal of Social and Clinical Psychology*, 17, 333-334.

O'Neal, E.C., Brunault, M.A., Carifio, M.S., Tronturne, R., & Epstein, J. (1980). Effect of insult upon personal space preferences. *Journal of Nonverbal Behavior*, 5, 56-62.

Osmond, H. (1957). Function as the basis of psychiatric ward design. *Mental Hospitals,* 8, 23-30.

Patterson, M.L. (1975). Personal space: Time to burst the bubble? *Man-Environment Systems,* 5, 67.

Patterson, M.L., Mullens, S., & Romano, J. (1971). Compensatory reactions to spatial intrusion. *Sociometry,* 34, 114-121.

Proshansky, H.M., Ittelson, W.H., & Rivlin, L.G. (1970). *Environmental Psychology: Man and his Physical Setting.* New York: Holt, Rinehart and Winston.

Reid, E., & Novak, P. (1975). Personal space: An unobtrusive measures study. *Bulletin of Psychonomic Society,* 5, 265-266.

Rumsey, N., Bull, R., & Gahngan, D. (1982). The effects of facial disfigurement on the proxemic behavior of the general public. *Journal of Applied Social Psychology,* 12, 137-150.

Schneider, F.W., & Hansvick, C.L. (1977). Gaze and distance as a function of changes in interpersonal gaze. *Social Behavior and Personality,* 5, 49-53.

Severy, L., Forsyth, D.R., & Wagner, P.J. (1979). A multi-method assessment of personal space development in female and male, black and white children. *Journal of Non-verbal Behavior,* 4, 68-86.

Shaffer, D.R., & Sadowski, C. (1975). This table is mine: Respect for marked bar room tables as a function of gender of spatial marker and desirability of locale. *Sociometry,* 38, 408-419.

Smetana, J., Bridgeman, D.L., & Bridgeman, B. (1978). A field of study of interpersonal distance in early childhood. *Personality and Social Psychology Bulletin,* 4, 309-313.

Smith, P.B., & Bond, M.H. (1993). *Social Psychology Across Cultural Analysis and Perspectives.* London: Harvester/Wheatsheaf.

Smith, R.J., & Knowles, E.S. (1979). Affective and cognitive mediators to reactions to spatial invasions. *Journal of Experimental Social Psychology,* 15, 437-452.

Smith, M.J., Reinheimer, R.E., & Gabbard-Alley, A. (1981). Crowding, task performance, and communication interaction in youth and old age. *Human Communication Research,* 7, 259-272.

Sommer, R. (1959). Studies in personal space. *Sociometry,* 22, 247-260.

Sommer, R. (1967). Sociofugal space. *American Journal of Sociology,* 72, 654-659.

Sommer, R. (1969). *Personal Space: The Behavioral Basis of Design.* Englewood Cliffs, NJ: Prentice-Hall.

Sommer, R., & Becker, F.D. (1969). Territorial defense and the good neighbor. *Journal of Personality and Social Psychology,* 11, 85-92.

Stokols, D. (1976). The experience of crowding in primary and secondary environments. *Environment and Behavior,* 8, 49-86.

Sundstrom, E., & Altman, I. (1976). Interpersonal relationships and personal space: Research review and theoretical model. *Human Ecology,* 4, 47-67.

Taylor, R.B., & Stough, R. (1978). Territorial cognition: Assessing Altman's typology. *Journal of Personality and Social Psychology*, 36, 418-423.

Tedesco, J.F., & Fromme, D.K. (1974). Cooperation, competition and personal space. *Sociometry*, 37, 116-121.

Tennis, G.H., & Dabbs, J.M. (1975). Sex, setting and personal space: First grade through college. *Sociometry*, 38, 385-394.

Thompson, D.E., & Aiello, J.R. (1981). *Non-verbal Behavior: Development in Retarded and Non-Retarded Children and Adolescents*. Paper presented at the meeting of the Society for Research in Child Development, Boston.

Ugwuegbu, D.C., & Anusiem, A.U. (1982). Effects of stress on interpersonal distance in a simulated interview situation. *Journal of Social Psychology*, 116, 3-7.

Watson, O.M. (1970). *Proxemic Behavior: A Cross- Cultural Study*. The Hague, Netherlands: Mouton.

Watson, O.M., & Graves, T.D. (1966). Quantitative research in proxemic behavior. *American Anthropologist*, 68, 971-985.

Wolfgang, J., & Wolfgang, A. (1971). Explanation of attitudes via physical interpersonal distance towards the obese, drug users, homosexuals, police, and other marginal figures. *Journal of Clinical Psychology*, 27, 510-512.

Worthington, M.E. (1974). Personal space as a function of the stigma effect. *Environment and Behavior*, 6, 289-294.

5

CROWDING

Most of the developing and under-developed countries including India are facing problems pertaining to the phenomenon growth in the population and overcrowding. In India, the population has already crossed one billion, making the country one of the most crowded in the world. This unabated increase in the population is partly responsible for various environmental problems like unhygienic living conditions, noise, and air and water pollution. A number of scholars, journalists and environmental activists have consistently noted that these environmental problems are presumably contributing to various societal, psychological and physical problems which mankind is facing today. For instance, Stokols (1976) has noted that "at the societal level mounting pressures from global problems like over population (Ehrlich & Ehrlich, 1970; Hauser & Dunken, 1959; Thompson, 1953), urban stress (Jacobs, 1961; Milgram, 1970; Simmel, 1950) and environmental deterioration (Carson, 1962; Commoner, 1963) have contributed to an increasing concern among policy makers, planners to layman about man's relationship to the physical environment."

During the 1960s, the concern of rising population and environmental deterioration has increased. With the publication of Ehrlich and Ehrlich (1970) most popular book entitled *Population, Resource and Environment* the idea of eco-catastrophe was born. The Ehrlichs argue that the world is already overpopulated in terms of world's ecological resources. They consider it possible that the capacity of the planet to support the human life in the planet earth has already been permanently impaired. "Spaceship earth" is now unable to sustain its over populated population and is slowly going out of control. In fact it has been reasoned by several scholars that since human being multiply at an exponential rate while goods and resources can only expand arithmetically, shortage, inflation and imbalance have become cancerous for our economy.

Keeping the abovementioned forecast and looking towards the present population trends, urban stress, resources scarcity, stress and its adverse affects on individual's quality of life, well-being and health, crowding which is a direct resultant of population appears to be an important area of research investigation. In this chapter attempt will be made to present a brief historical background of crowding research followed with the conceptualization of the phenomenon of crowding and how it is different from density. Furthermore a comprehensive review of crowding studies will be presented in diverse settings. Finally, the theoretical perspectives and future direction of crowding research will be presented.

Historical Background

Three important traditions have historically dominated crowding research in Euro-American context. These are:

1. In 1920s the sociological tradition, which centred on urban analysis was popular.
2. In 1950s and 1960s studies on animal crowding was the dominant focus.
3. In the late 1960s and early 1970s the socio-psychological trend was emphasized.

A brief sketch of these three perspectives is given below:

The Sociological Tradition

The early urban sociologists concentrated on correlational methods and tried to study the relationship between population density and various outcome measures related to social pathology, like juvenile delinquency, physical and mental health. Another characteristics of this research were its emphasis on long-term outcomes such as crime rates, mental and physical illness. Finally, the effects of crowding were studied primarily as a societal level problem and not as an individual, family or small group level.

The Animal Ethnological Tradition

Research on animals began in 1960s. The foremost advantage of animal research is that one can telescope in time the impact of population density because in comparison to humans the life expectancy of animals is shorter and thus social processes and reactions to crowding can be analyzed more comprehensively. In addition, detailed autopsy and variation of population density in various experimentally created conditions is possible which is not possible when human subjects are involved. The major limitation of these studies is that generalization from animal to human subjects is not always possible.

The Socio-Psychological Tradition

In the early 1970s the research of crowding on humans began. Early research on crowding was conducted primarily in the laboratory settings with an emphasis on control of extraneous variables. Groups were formed under varied conditions of crowding and its effects were studied on task performance and feelings. However, after couple of years research on crowding has gone into high gears and the effects of crowding has been examined in different field settings including home, cities and prisons.

Conceptualization of Crowding

To a layman on the street the meaning of crowding is often confused with density. It was believed that crowding is simply having "too many people at some place at some time" and the result is the negative feelings and reactions towards others. But this view does not reveal much and some pertinent questions remained unanswered. For instance, how many people, how close together the people are and how long a time people are exposed to such noxious conditions.

There is no all-inclusive definition of crowding, however, to reach at a more comprehensive definition one has to have a close look on some of the frequently used definitions of crowding. For instance, Proshansky, Ittelson, and Rivlin (1970) proposed that "crowding is not simply a matter of density of persons in a given space. The experience of "being crowded" depends to some degree on the people crowding him, the activities going on and his previous experience involving number of people in similar situations." Stokols (1976) had made a distinction between the physical condition of density involving spatial limitation and the experiential state crowding in which the restrictive aspects of limited space as perceived by the individual exposed. Sometimes "crowding" is viewed in terms of spatial consideration alone. But this meaning of crowding fails to delineate those social and personal aspects, which may interact with spatial factors in the experience of crowding. According to the proposed definition of Stokols (1976) "density is viewed as a necessary antecedent, rather than a sufficient condition for the experience of crowding". According to Desor (1972) "Being crowded" is receiving excessive stimulation from social sources. Desor on the basis of his research suggested that although judgment of crowding varied with space, the experience of crowding is much more than a matter of space. Though, level of social stimulation appears to control judgments of crowding when other things are held constant. It seems that those 'other things' are important in deciding when the situation is crowded. Crowding may be considered as a psychological concept referring to a subjective experience, which may or may not be adequately reflected by population density measures such as amount of space per person or number of person per unit, or number of people per unit of living space (Baron, Mandel, Adams, & Griffitt, 1976). Schmidt and Keating (1979) defines that crowding is experienced when situational density forces the blocking of goals, interruption of behaviour or cognitive overload to appear. According to Epstein and Baum (1978) crowding involves an imbalance between people and available space. There seems to be three events that occur as a result of this imbalance: Congestion resource scarcity, an ability to control and limit interaction with others and extremely close proximity to others.

In addition to the above mentioned operational definitions about crowding, some frequently used conceptualization focuses on amount of space available (Freedman, 1975; Stokols, 1976), close physical proximity (Aiello, Epstein, & Karlin, 1975; Worchel & Teddlie, 1976), reported crowding (Jain, 1997), unwanted interaction and overload (Baum & Valins, 1977; Saegert, 1978), inability to control interaction with others (Zlutnick & Altman, 1972). Finally, based on the contemporary theoretical models, some attempts have been made to delineate subjective parameters of crowding. Factor analysis results revealed evidence of four distinct crowding experience components labelled as: spatial satisfaction, positive or supportive relationship, negative or disruptive relationship and uncontrolled disturbance (Nagar & Paulus, 1997). Thus, it seems that conceptualization of crowding involves multifaceted experience of individual in response to density related factors. There are a number of variables that will influence our experience of crowding. These will include our relationship with the people involved, the duration of the experience, the physical context of the experience, and the meaning of the experience (Cassidy, 1997). We are likely to feel crowded if the place is occupied with strangers relative to that of friends. We have also experienced positive emotions when the theatre is occupied to its capacity. In contrast, if most of the seats of the theater are empty we fail to enjoy low density. The importance of the physical and social context together with the degree of choice seems to be important in ascertaining the crowding experience.

Review of Crowding Literature

A number of researchers have comprehensively reviewed the crowding literature (Baum & Paulus, 1987; Jain, 1997; Nagar,

1998; Pandey & Nagar, 1987). It seems that research on crowding can be grouped into three broad categories:

1. Animal Studies
2. Correlational studies
3. Experimental studies

Animal Studies

Early research on crowding and high population density was done almost exclusively on animals (Calhoun, 1962; Christian, 1963; Mardson, 1972; Lawrence, 1974). A series of studies conducted by Calhoun (1962, 1973) are pertinent to the effects of crowding and social behaviour of rats. He confined the rats in a quarter of acre enclosure. Group of rats in the laboratory confinement were allowed to breed freely and were supplied with adequate food, water and nesting material. As the population of rats grew in size the adverse effects of crowding was observed. In fact the behaviour of rats under crowded conditions deteriorated so much that Calhoun coined the phrase behavioural sink to describe the adverse effects. One very important finding was that by the end of twenty-seven months the population has become stabilized at a particular level. In fact the population levelled off at 150 while the space was sufficient to accommodate several thousands. The reason for this was very high level of infant mortality caused by aggressive attacks from adult males on pregnant females and often cannibalism in eating the newborn. In general there appeared to be increase in psychopathology. The common symptoms associated with high density, which served ultimately to reduce the density of the colony included catatonic withdrawal, sexual aberration, disruptive reproductive functions and aggression. In 1973 Calhoun replicated his earlier study conducted in 1962. The results revealed that despite adequate facilities to accommodate 4000 rats the population never crossed 2200. Disruption of reproductive processes resulted in the complete extinction of the rat colony.

One of the another frequently cited field study concerning density and animal behaviour was conducted on a small herd of silka deer on the island of Chesapeake bay by Christian, Flyger & Davis (1960). The food, water and other resources were in abundance and the deer were left free to breed. The herd increased in number until density increased the ratio of one deer per acre. At this point mortality rate increased. Subsequent examination of carcasses revealed a variety of endocrinological disorders apparently resulting from extreme stress brought on as a reaction to crowding symptoms including enlarged adrenal glands by fifty per cent and reproductive dysfunctions. Deevey (1960) reported increased mortality in lemmings and snowshoe hare. A similar trend of results with regard to high density and adverse consequences were reported on cats (Leyhausen, 1965) and monkeys (Southwick, 1967).

There are some studies conducted on animals in which the relationship between crowding and index of performance was seen. Goeckner, Greenough and Meed (1973) in their experiment found that rats reared under crowded conditions were impaired in learning brightness discrimination tasks to avoid shock as compared to previously uncrowded animals. The adult rats who were previously isolated for 35 to 90 days when tested again after spending four weeks under high density conditions were superior to those previously isolated rats of low density conditions. Similarly, in Leshley maze a tendency towards facilitation of performance task was poor in crowded adult group in contrast to young crowded reared rats.

In conclusion, it seems clear that crowding is an aversive situation for animals, which disturbs the normal social behaviour, affects the physiology and leads to various pathological behaviour. However, there are many problems in animal studies in terms of both ethics involved and generalization of findings for human subjects.

Correlational Studies

Guided by the early research on animals social scientists

became interested to test the effects of crowding on human. The strategy used generally by many investigators was aimed to examine the relationship between various indices of population density with social pathologies. The strategy generally has been to correlate measures of density like population per acre, number of persons per room, number of dwellings per unit of space and so forth with various measures of pathologies including crime rates, mortality rates, mental and physical health data and suicide rates.

Evidence exist that high population density and crowding results in various types of psychological and social disturbances and disorders. Zlutnick and Altman (1972) have mentioned broadly three categories of evils in response to crowding. These are:

1. *Physical effects:* Starvation, pollution, slums, disease, physical malfunctions...,
2. *Social effects:* Poor education, poor physical, mental and health facilities, crime, riots and war...,
3. *Interpersonal and psychological effects:* Family disorganization, addiction, aggression, decreased quality of life (p. 49).

Correlational studies can be broadly divided into two broad categories:

1. Studies on Cities,
2. Residential Crowding

Studies on Cities

From the frequent influx from rural to urban areas cities are now crowded beyond capacity. For many people life in the city today is considered as unpleasant, dehumanizing and dangerous. Furthermore, by the insufferable living conditions cities are rapidly changing from bad to worst for many people. Investigations of demographical data collected on city population in Hong Kong (Schmitt, 1963), Honolulu (Schmitt,

1966), and Chicago (Galle, Gove, & McPherson, 1972; Winsborough, 1965) revealed lack of significant correlation between density and social pathology. However, the results of these studies provide some indication of positive relationship between higher crime rate and increased density. To study the links between crowding and crime Gifford and Peacock (1979) conducted a study in Toronto and Hong Kong. The results reveal that population of Hong Kong was four times more than Toronto, yet the crime rate was approximately four times less in Honk Kong relative to Toronto. Crowding is associated with increased levels of aggression and violent behaviour and may be a contributing factor in assaults on spouse or children in family settings although the evidence is inconclusive (Fuller *et al.*, 1993). Thus, it is important to examine the links between crowding and specific crimes. For instance, crimes like pick-pocketing tend to occur more frequently in crowded situations. Marsella, Escudero and Gordon (1970) studied the relationship between dwelling density and mental disorders in males in Manila. They found that with the increase in population density the indices of mental disorders such as psychosomatic complains, alienation, anxiety, withdrawal, eruptive violence and "free floating anxiety" were found to increase.

A study was conducted by Schmitt (1966) to examine the relationship between density and various pathologies in Honolulu. He found a strong positive correlation between population per residential acre and nine measures of health and well-being. Schmitt reported relationships between density and death rate, incidence of tuberculosis and venereal disease, mental hospital admissions, juvenile delinquency, and imprisonment. Even after controlling the income and educational level, partial correlation was found highest with average rates of venereal disease, first admission to mental hospital, and lowest with infantile mortality and suicide. Winsborough (1965) conducted a study in Chicago to investigate the relationship between population density with health and well-being. After controlling the effects of socio-economic status, quality of housing and ethnicity, he found that mortality, tuberculosis, and general public assistance were negatively correlated with density, while correlations with infantile mortality and public assistance to minor were positive in nature. Booth and Johnson (1973) after controlling for income and education reported that density was related to the rates of number of different crimes, including assault, rape, robbery, burglary, theft and larceny. Similar results were obtained by Loo (1973) in a demographical survey of mental health trends in Hong Kong. He found that with expanding population and urbanization over a period of twenty-five years, the incidence of mental illness increased. In addition, no adaptation to high-density living conditions was found.

Galle *et al.* (1972) conducted a study in 75 community areas of Chicago. He used four separate measures of density—person per room, room per dwelling, dwellings per structure and structures per acre. A combination of these four densities independent of social class and ethnicity, accounted for substantial portion of variance in the rates of mortality (23%), juvenile delinquency (25%), fertility (14%), public assistance to minor (34%) and admission to mental hospital (26%). Furthermore, inside density was related to crime and health variables while outside density was not. Furthermore, in a multinational study Booth and Welch (1973) found that household density appeared to be related to more indicators of pathology than did outside density.

A number of studies were conducted to examine the relationship between crowding and suicide. Carstairs (1959) reported that while suicide is related to social isolation, attempted suicide is associated with high-density condition. In Santiago, Chile positive correlation was found between crowding and suicide (Chuaqui, Lemkau, Legarrefa, & Contresas, 1966), but Sainsbury (1955) in London and Lester (1970) in Buffalo, New York failed to find any relationship between density and suicide. Cassell (1971) reported that the incidence of suicide is inconsistently related to population density. Schwab, Nadeau, and Warheit (1979) conducted an epidemiological survey on the mental needs and services of residents living in Southeastern part of USA. Results indicated that crowded subjects scores significantly higher than uncrowded subjects on a depression scale. Associations between crowding and high scores on both scales were strongest among (a) Subjects in the child rearing and middle years of life, (b) Blacks at all income levels, (c) Whites in the intermediate family income levels, and (d) Females rather than males. The crowded black and the crowded white population reported more depression relative to their uncrowded counterparts. Thus, it is revealed that the three variables being a female, having a low income and crowding accounted for more than sixteen percentage of variance. In crowded situation, depression may be a costly semi-adaptive reaction to excessive interpersonal stimulation. Women living in crowded situations appear to be at high risk for depressive illness, their plight is reminiscent of animal subjects (Calhoun, 1962) indicating that the maternal behaviour of females deteriorated in crowded situations.

Thus, the overall pattern of results is rather mixed and provides only weak evidence of a relationship between urban density and pathology. Furthermore, several studies demonstrated evidence that living alone is positively related to increased use of stress-reducing drugs, to suicides, and to admissions to mental hospitals (Levy & Herzog, 1974; Galle *et al.*, 1972). While the correlational studies are having its own limitations (Baum & Paulus, 1987; Epstein & Baum, 1978), they do provide information about the potential generality of the results from field studies.

Residential Crowding

Most of the studies have reported that control and group orientation are shown to mediate reactions to high density in a variety of residential settings. The literature on residential crowding can be broadly divided into three parts:

1. Crowding in Dormitories
2. Home Crowding
3. Crowding in Prisons

Crowding in Dormitories

The dormitory setting is mainly individualistic in character and it ranges from cooperative to competitive according to variation of the social situation. MacDonald and Oden (1973) in their study demonstrated that youth living in crowded dormitory style accommodations with perception of high degree of control and strong sense of cooperation reported positive feelings. In this study five married couples, who have volunteered to share an unpartitioned 30 by 30 foot room for 12 weeks of training were compared with volunteer couple that lived in hostel accommodation. The crowded couples not only failed to show adverse affects but also showed enhanced marital relationships. It seems that these couples have developed a high degree of cooperation with each other and high degree of control over the environment and thus crowding did not adversely affected them.

The effect of residential crowding was examined by comparing the experiences of tenants living in 14-storied apartment buildings and those living in 3-storied, within a low-income housing project. The results obtained by McCarthy and Saegert (1978) states that since tenants in high rise buildings would come into contact with large number of others in the public spaces of the buildings and as these unwanted contacts exceeds residential interaction capacity, tenants would experience social overload. This experience would be manifested by tenants perception of crowding in the building, feelings of less control, safety and privacy in their immediate environment, problematic social relationship among tenants, alienation and dissatisfaction relative to their counterparts living in 3-storied apartments. In addition, to this high rise apartment residents were less socially active beyond their building and felt greater degree of powerlessness in affecting management decisions.

In a series of studies Baum and his associates examined the effects of two different architectural designs on social behaviour and experience of crowding (Baum, Aiello, & Calesnick, 1978; Baum & Valins, 1977; 1979; Baum & Gatchel,

1980). A comparison was made between groups of students living in traditional double loaded corridor dorms with students residing in suit style accommodation. In other studies long corridor dorm students were compared with students living in short corridor dorms. All residents engaged more or less equal square footage of living space. It was found that because long corridor residents were more likely to encounter unwanted social interaction they reported more lack of control over the environment and less feeling of cooperation than their counterparts. Furthermore, it was reported that long corridor residents felt more helpless and acted more competitively than students living in accommodation with different architectural variety. In addition, such students perceived high degree of crowding and experienced more stress than their counterparts.

The studies conducted by Baum and Valins (1977, 1979) attempts to focus on the relationship between social density and problems faced by dormitory residents while regulating their social interaction. The residents residing in long corridors shared their bedroom with single other person, and the hallways, bathroom and lounge with 33 others. In contrast, short corridor residents shared their room with single others and other amenities with 18 or 19 residents. Finally, the suit residents shared their bedroom with single other but shared the remaining amenities like lounge and bathroom with 3 or 5 others and only the hallway with 33 neighbours. Thus, the functional social density was highest in case of long corridor dorm residents followed by short corridor and the lowest for suit design dorm residents. Buildings of long corridor dorms were associated with reports of excessive social stimulation and unwanted social contact with acquaintances and strangers in the common areas of the floor leading to crowding stress. In contrast, residents of short corridor and suit dorms were able to develop mechanism of regulating social interaction because the effective group size was small. The small group size helped the residents to exert more control over the environment and the reduction of unpredictability was reported. Residents became familiar with one another because the group size was small, which allowed the formation of norms and other ways of regulating interaction with others. Furthermore, besides self-report, extensive behavioural data were collected and behaviour of the residents was observed both in the dorms and laboratory settings. Long corridor residents reported elevated feelings of crowding, complained about their neighbours and unwanted interactions and stayed isolated in their bedrooms for a longer period of time relatively to their counterparts. This avoidance tendency reported by the long corridor residents persisted even in other settings. The residents residing in long-corridor dorms initially reacted to loss of control over social experience by trying to reassert control over their interactions (Baum, Aiello, & Calesnick, 1978; Baum and Gatchel, 1980). After 5 or more weeks' failure to achieve control appear to diminish controlling behaviour and allow the emergency of withdrawal reactions. The results are supported by many scholars (Wortman & Brehm, 1975) and can be explained within the framework of learned helplessness (Seligman, 1975).

Karlin, Epstein and Aiello (1978) conducted a study at Rutgers University and compared the students who lived in triple and double loaded corridors. It was found that tripled students were more disturbed, disappointed and stressed than doubled students. These results were more severe in tripled women. The reason behind seems that women spend more time in their rooms in their attempt to make home like environment. In contrast, men spend more time outside their dorms. A survey was conducted by Aiello, Baum and Gormley (1981) to examine the experience in the dorms and adjustment to college life. The results of the survey revealed that triples as compared to doubles showed greater social tension, negative affect and control problems. Furthermore, they found that these symptoms of interpersonal stress are mediated by group formation within tripled rooms. Isolates left out by a coalition of other room mates were more vulnerably to crowding and experienced more control related problems. Similar results were obtained by other scholars (Baron, Mandel, Adams, & Griffen, 1976; Reddy, Baum, Fleming, & Aiello, 1981). For instance, Reddy *et al.* (1981) found that isolates in both three or four person groups reported more crowding, more complains related to dissatisfaction with room and room mates, difficulty in maintaining control, and less confidence in their ability to regulate social interaction with non-isolates. However, tripled isolates consistently reported significantly more complains regarding not been understood by their room mates and experienced loneliness, as well as greater exclusion from group decisions and difficulty in regulating social interaction and control related problems. Finally, it was also reported that when the group size increased from three to four, crowding stress did not increase.

Indwelling Crowding

The indwelling crowding is conceptualized as groups of people who are normatively expected to be cooperative and interdependent. In order to investigate the effects of crowding in the family dwelling two types of strategies have been used. Prior to 1970s, the investigators relied on the ecological correlations. But this methodology was severely criticized (Epstein & Baum, 1978) for it failed to provide any clear and reliable evidence that crowding and pathology are related or not. Freedman (1975) found no relationship between crowding and pathology. But Johnson and Matross (1977) found a clear relationship between density and psychiatric hospitalization. Thus, it can be speculated that adverse effects of crowding are more pronounced in people who are mentally ill, are being constrained by their individual needs, and are less likely to cooperate with other family members. Cross-sectional and longitudinal studies have shown that residential crowding can adversely affect psychological health (Evans, Palsane, Lepore, & Martin, 1989; Lepore, Evans, & Schneider, 1991).

The second type of studies utilized interviews with residents. Mitchell (1971) conducted a study in Hong Kong. The occupant shared a 400 square foot dwelling space with 10 or more persons. This averages out to be 40 or less which is less than half of the 85 square foot per person considered a minimum in households in which all occupants are normally healthy according to Committee of the Hygiene of housing, American public health association (1950). No evidence of deficits in emotional health was found when density was high. A significant interaction effect between the presence of unrelated family within the household and the floor level on which the house was located was found on the dimension of emotional health. The persons living on the sixth floor or above in households consisting two or more unrelated families showed a greater degree of emotional illness and highest level of hostility. One of the major reasons seems to be the disruptive social relationship between occupants living in crowded residential homes. However, adverse affect of chronic crowding may result in social withdrawal, which is a coping strategy people adopt to deal with crowding (Lepore, Merritt, Kawasaki & Mancuso, 1990).

A study was conducted in Toranto by Booth and Johnson (1975) utilized both objective and subjective measures of crowding. The results indicated no major differences with regard to variation of density. However, it was found that men who grew up in crowded households were less likely to experience stress related disease under conditions of high density than men who grew up in less crowded households. Gove, Hudges and Galle (1979) conducted their study in Chicago. They found that both objective crowding as measured in terms of person by room ratio and subjective crowding as measured by excessive social demands and lack of privacy were strongly related to poor mental health, poor social relationship at home and poor child care, and less strongly but significantly related to poor physical health and poor social relationship outside the home. The study further reports that overcrowding in the home is associated with irritability, withdrawal, weariness, ineffectual behaviour, and a variety of indicators of poor mental health. This trend of result has received support from many researchers (Loring, 1955). While summarizing their results Gove *et al.* (1979) reported that in crowded households children are experienced as irritants and it is a relief when they are out of their home. In such households parents are less informed about their children activities outside their homes.

Crowding in Prisons

Prisons provide a useful naturalistic site for conducting studies on crowding because of various reasons including the availability of wide range of housing conditions and long duration of exposure to high-density conditions. In contrast, the studies conducted in laboratory setting are typically brief. One of the most crowded chronic conditions in household setting that has been studied involved up to 9 people living in one bedroom (Rodin, 1976). The conditions prevailing in a typical prison setting is more chronic in terms of crowding and density. For instance, researchers study prison conditions in which up to 70 inmates are confined in open dormitories with less than 20 sq. ft per person. A 1000 sq. ft apartment would have to house 50 people in order to match this level of spatial density alone, and social density in prison settings was also considerably higher than has been typically examined. Paulus, McCain and Cox (1981) in their study found that high levels of density in prison setting is sufficient to produce indications of pathology. In addition, McCain, Cox and Paulus (1976) found that prisoners sharing a dormitory cell with others had higher numbers of illness complains than those housed in a single cell. D'Atri (1975) in their study on prison crowding compared blood pressure of inmates living in dormitory type cell with those housed in single cell. The results indicated that both systolic and diastolic blood pressure was higher for inmates sharing a cell. In another study Paulus, McCain and Cox (1978) by way of archival data tried to examine the relationship between crowding in prisons and various measures of pathology. They found that more crowded the prisoners were, the higher were the death rates, the number of psychiatric complains, and blood pressure. In another study, Cox, Paulus and McCain and Schkade (1979) found a positive correlation between levels of density and palmer sweat of prisoners.

The study conducted by D'Atri (1975) investigated the effects of crowded housing on blood pressure in 568 male prisoners as an analog to animal studies. The results indicated that crowding elevates blood pressure. The transfer from single occupancy cells to multiple occupancy dorms was associated with a statistically significant mean difference in systolic blood pressure. Subjects who lived in single cells had little mean change in systolic blood pressure over time. On the other hand the subjects who were retransferred to cells after a short stay in the dormitories experienced a mean decline in systolic blood pressure, which suggest, that effects of crowding may be reversible in its early stages. Furthermore, it is also seen that blood pressure decreases after continued stay in the dormitories, indicating that adaptation may occur. Based on a series of studies conducted by Paulus and colleagues, it was concluded that the primary cause of negative effects related to crowding may be due to cognitive strain, anxiety or fear and frustration intrinsic to most social interactions in crowded settings (Paulus, 1988). In Texas, where almost all of the prisons are maximum security prisons, large prisons (Pop, 1400; Mean 1700) were found to have 10 times more suicide rates, 78 per cent higher psychiatric complain rates, and three times higher death rates than small prisons (Pop, 1100; Mean, 830). (Paulus *et al.*, 1981). Furthermore, crowding was also found to be associated with disciplinary infractions and suicides. The prison crowding results based on archival data should be interpreted carefully because many confounding variables operate simultaneously and their effects on dependent variables are likely to contaminate the obtained trend of results.

Experimental Studies on Crowding

The primary aim of experimentation is to increase control over extraneous variables and the possibility of establishing causal relationships between dependent and independent variables. A large number of studies were conducted in the laboratory setting in which crowding is usually engendered by placing varied number of individuals in a small space for short period of time. A brief review of laboratory studies on crowding is presented below:

Physiological Effects of Crowding

A large number of studies have examined physiological response to high density have considered peripheral measures associated with stress and arousal. The duration of exposure to conditions of high density in the experimental setting is usually very short. However, even brief encounters (30 to 75 minutes) with high density can affect physiological responses including skin conductance and increase in perspiration (Aiello, Epstein & Karlin, 1975; Saegert, 1974). The study of Evans (1979) attempts to examine the physiological responses after a period of three hours of exposure to high-density conditions. The result indicated elevation of blood pressure in response to crowding. Some of the quasi-experimental field studies related crowding with endocrine response (Lundberg, 1976; Singer, Lundberg & Frankenhaeuser,1978). These studies were conducted on respondents travelling in Swedish commuter train under varying levels of density. The analysis of the urine samples collected from the respondents after the termination of trips revealed higher levels of epinephrine among subjects after crowded trips relative to less crowded trips. Incidentally, epinephrine is associated with other measures that have been used to measure stress including blood pressure, skin conductance, heat rate (Baum, Grunberg, & Singer, 1982). Some researchers have suggested specific aspect of crowding in commuter trains. Regardless of the level of density passengers boarding the train at the first stop experienced less distressed and exhibited lower levels of epinephrine than their counterparts who boarded the train in the halfway. It seems that passengers who boarded the train at the first stop were able to select their seats and their partners in the empty train. The control afforded these first boarders may have reduced the adverse effects of increasing numbers of riders along the way (Lundberg, 1976; Singer *et al.*, 1978).

A number of researchers examined the links between crowding and illness in various residential settings including prisons and dorms. For instance, McCain, Cox, and Paulus (1976) reported that prison inmates living under low-density conditions made fewer illness complains relative to their counterparts exposed to high densities. Further studies indicated that density was related to increase in blood pressure and death rates (Paulus, McCain & Cox, 1978). Similar results were obtained by other scholars in prison settings (D'Atri, 1975; D'Atri & Ostfeld, 1975). In a similar vein Stokols and Ohlig (1975) reported that crowding in the college dorms was associated with visits to the students health care centre. Available evidence in diverse settings suggests that crowding and increased density is associated with physiological and psychological arousal and is linked with increased skin conductance, blood pressure, endocrine activity and illness complains.

Crowding and Task Performance

Before 1975, a number of studies examining the effects of high density in the laboratory situations had failed to find any effects on task performance (Freedman, Klevansky, & Ehrlich, 1971; Bergman, 1971; Stokols, Rall, Pinner, & Schopler, 1973). For instance, the study conducted by Freedman *et al.* (1971) is given special attention. They have reported the findings of three experimental studies in which groups of people performed tasks under varying conditions of density. The tasks varied in complexity. In all cases the subjects worked alone, in the presence of others in groups of five to nine, in varied available spaces from thirty-five to one hundred and sixty square feet. The investigators found no effects of any kind attributable to the degree of population density. However, other studies did report performance deterioration in response to crowding (Dooley, 1974; Saegert, 1974).

Most of the research conducted after 1975 more strongly suggested that crowding can impair complex task performance rather than simple ones (Aiello *et al.* 1975; Evans, 1979; Paulus, Annis, Seta, Schkade, & Matthews, 1976; Nagar & Pandey, 1987). Most of the tasks used involved problem solving, discrimination, and concentration and story comprehension. The results obtained in these studies suggested that crowding effects are most likely on complex task performance. Furthermore, certain features of the task situations like perceived control and the use of multiple tasks seem to influence the degree to which crowding effects are observed (Paulus & Matthews, 1988). The results found in one simulated experiment conducted on five hundred subjects ranged from elementary school age to elderly by Karlin, Epstein and Aiello (1978) indicated that when compared to their non-crowded counterparts, subjects working in crowded conditions showed increase in simple and decrease in complex task performance. Another important study was conducted by Sherrod (1974) in which the experimenter asked his subjects to perform simple and complex tasks both during and after exposure to high density under different conditions of perceived control. The results of this study revealed that performance of both simple and complex task was unaffected as a result of density variation. However, on post-experimental tasks performance decrement was observed in subjects who were not given perception of control of the stressors.

Klein and Harris (1979) utilized a 2 (high *versus* low density)* 2 (simple *versus* complex) * 2 (forwarding *versus* non-warning of crowded room) design to examine the effects of crowding on task performance on undergraduate students. The results of this study indicated that more tasks were attempted and efficiency was higher when forewarning about crowd was given to subjects relative to the conditions where subjects were not warned about the crowd. Cohen (1978) conducted an experiment on undergraduates and found that crowded subjects relative to their counterparts who worked under less crowded conditions showed a moderate decrement in complex task performance. The performance of simple task was not affected due to the variation in the degree of crowding. Furthermore, poorer performance in a group cooperation task and less tolerance for frustration as an after-effect was reported by crowded subjects. Evans (1975) reported that when subjects worked on two tasks at once, performance on secondary task suffered in high-density conditions. Crowded subjects relative to uncrowded subjects exhibited elevation in blood pressure and pulse rate and manifested additional behavioural indices of stress, including greater stereotypes, increased defensive posturing, more remark toward the door of the room and self-report of greater hostility (Evans, 1979).

Social Response to Crowding

Crowding is essentially a social phenomenon primarily involving nature and degree of interpersonal contact and behaviour. Two basic forms of social responses, namely withdrawal and aggression are identified by researchers. According to Baum and Paulus (1987) withdrawal or avoidance of social contact has been found in settings where too many people are present and reported crowding is related to frequency or quality of social contact. The aggressive response has been observed in situation characterized by fewer people working in very small spaces. In contrast, the reported crowding is due to spatial restriction and inappropriate physical proximity with others.

A number of studies conducted in residential setting have reported evidence of withdrawal, which is considered as an active coping response to deal with unwanted interaction with large number of people (Baum & Valins, 1977). The expectation of being in a room with a large number of people also elicits withdrawal tendencies in the respondents including reduced level of eye contact, increased inter-individual distances, and head movements away from others (Baum & Greenberg, 1975; Baum & Koman, 1976). In high density residential areas people disregard low priority inputs (e.g. strangers, casual acquaintances) sometimes through gestures and postures and allocate attention in favour of more important contacts including friends and relatives. Eye contact is an initiator of social interaction and looking away is considered as withdrawal response. In small towns willingness to interact with strangers was greater relative to big cities and urban areas (McCauley & Newman, 1977). In one study it was found that people show withdrawal tendencies as an active attempt to cope in high density situations (McCauley, Coleman, &

Defusco, 1977). Furthermore, people are less likely to extent help to strangers in emergency in urban areas is amply reported in the literature. In a similar vein, a number of studies have reported that people are less likely to help one others when they feel crowded (Korte, 1979; Jorgenson & Dukes, 1976).

One way of coping with crowding is to withdraw from social interaction and make an attempt to restore some sense of control over the situation. Evans and Lepore (1992) in a simulated experimental study compared participants from crowded living conditions with participants from uncrowded homes. They found that respondents residing in crowded homes were less likely to offer support or ask for support relative to their counterparts. Evans *et al.* (1989) reported that the effects of household crowding are mediated through its effect on social support with negative consequences occurring where the social support system within the family was disturbed. The study conducted by Ruback and Pandey (1992) found that participants who had higher levels of perceived control, were less affected by crowding. In another study Ruback and Patnaik (1989) found that crowding facilitates act of vandalism and destruction of property in situations where the degree of perceived control was low.

Crowding leads to increased physiological arousal and stress is reported in the environmental behaviour literature. Because of its effect on arousal, crowding instigates the process of emotional responding and could produce aggression and anger or apathy, social withdrawal or even helplessness. Baum and Paulus (1987) suggested that both withdrawal and aggressive behaviour may be a realistic approach to deal with crowding and high density situations. A number of studies have demonstrated that crowding leads to aggression (Aiello, Nicosia, & Thompson, 1979; Hutt & Vaizey, 1966). Several intervening conditions explain the nature of aggressive response in crowded conditions. For instance, Rohe and Patterson (1974) suggested that competition for scarce resources, including space may probably be one important consideration for the display of aggressiveness during conditions of high density. Some studies have reported gender differences as an important determinant of a crowding-aggression linkage. Studies have reported that men exhibit more aggressive responses to crowding relative to women during exposure to spatial restriction (Freedman *et al.*, 1972; Stokols *et al.*, 1973). Baum and Koman (1976) found evidence of aggressive posturing as a preparatory response when expecting crowding in a small room. Men anticipating interaction with four other person in a small room assumed more aggressive seating position and interaction styles than did males anticipating a session with the same size group in a larger room. In contrast, women did not exhibit aggressive responses. In case of larger group withdrawal tendency rather than aggressiveness was observed.

It seems clear from the research that people usually exhibit various coping strategies to deal with the adverse consequences of crowding. Some of the important mechanisms involved in dealing with adverse consequences of crowding are related to coping and social support. However, more studies are needed to determine the exact casual path of the complex linkages between mediators of crowding and ill effects.

Crowding and Adaptation

Glass and Singer (1972) defines adaptation as "a cognitive process involving one or other mechanism designed to filter out of awareness certain aspects of an aversive event or in some way reappraise it as benign." It is often seen that when an individual is exposed to stressors he develops certain strategies to reduce or eliminate the adverse impact. However, the results of various studies are inconsistent and one is not able to say with confidence that adaptation to crowding takes place or not. For instance, Sundstrom (1975) found no clear evidence of adaptation. However, the results revealed discomfort and crowding both in high and low-density conditions. However, decrease in crowding was accompanied by non-verbal coping behaviour and this response of the subjects to the density condition is successful coping and not adaptation. Furthermore, despite coping, irritation was increased over time because of goal blocking. The results obtained from the study conducted by Aiello *et al.* (1975) reveals that when college students were enclosed in a very small room in groups of six, their skin conductance increased over time, indicating the opposite of adaptation. It has been reported in some studies that people not only respond to dangers or threats that have materialized but mere anticipation of crowding for example, leads an individual to experience stress (Baum & Greenberg, 1975). Finally, in some studies it has been reported that aversive consequences persists even after the source of stress is terminated. In other studies, a post-stressor effect, which is opposite to adaptation, is seen for noise stressors (Glass & Singer, 1972) and for crowding (Evans, 1979; Sherrod, 1974). Thus, it is clear that crowding in the laboratory is perceived as a stressor. In some studies adaptation is found while in the other cases opposite of adaptation is reported.

Developmental Effects of Crowding

A number of studies suggest that adults relative to children experience crowding at lower density levels. The study conducted by Sinha, Nayyar, and Mukherjee (1995) utilized a wooden box, which was divided into two sections with a connecting door. Participants were asked to consider dolls as people and were instructed to populate the space. The participants comprised of both children and adults were asked to imagine that they were hosting a party. The results of this study indicated that children introduced more people than adults. Thus, it seems that as people grow older the personal space requirements also increases. Home crowding seems to negatively affect children more than adults. It seems that adults encourage their children to spend more time outside their homes (Gove *et al.*, 1979). The major results of the study conducted on black South African children reveal that parents' education, occupation and crowding are three major predictors of cognitive development and self-concept (Goduka, Poole, and Aotaki-Phenice, 1992). Crowding is also found to be a major predictor of physical underdevelopment in British children.

In a study Powell (1990) found high-density living and conditions of uncontrolled noise creates a psychological state of helplessness responsible for the deterioration the performance of African-American children in mathematics and science. Widmayer *et al.* (1990) reported that household crowding and low birth weight were potent predictors of psychomotor development of 12 months old children. Evidence exist that the mother-child interaction is adversely affected by household crowding. Researchers have reported that involvement and verbal stimulation given to the child is reduced and a tendency of frequent disciplining the child tends to increase with regard to increase in home crowding (Fuller, Edwards, Vorakitphokatorn, & Sermsri, 1993).

Based on the above mentioned findings it seems that crowding is one variable that coexist with many other variables embedded in the home environment. To draw any firm conclusion, it is essential to eliminate the effects of other variables present in the home environment. However, it is plausible to infer that crowding adversely affects the development process in children.

Individual and Cultural Difference to Crowding

Not everyone is affected by crowding in the same way. People residing in crowded countries or crowded homes show higher threshold of tolerance to crowding and other elements having similar properties. People are living comfortably for a prolonged period of time in high altitudes where supply of oxygen is low. This situation is very uncomfortable for people living in plains. It seems that people develop mechanism to deal with the adverse effects emanating from environmental challenges. Prior experience of high crowding in the residential settings prepares the individual with high reactivity in dealing with situations having the similar elements (Nagar *et al.*, 1988). Other studies have similarly reported that prior experience and expectations influence both our sensitivity to crowding and our coping responses (Lepore *et al.*, 1991; Webb & Worchel, 1993). With regard to cultural effects, some studies found that

Blacks as compared to Hispanics in Chicago (Gove & Hudges, 1983) and among foreign students relative to Indian students in India (Odera & Hasan, 1993) have more sensitivity to crowding. According to Cassidy (1997) studies conducted across cultures are very diverse to allow deriving firm conclusions. He advocates that in recent years because of high people mobility many countries contain mix of people from diverse cultures. It seems that people with prior experience of high crowding are likely to respond positively to reduce density in a new environmental setting. He feels that prior experience of crowding, expectations, and current commitments rather than culture is to be given more consideration.

Research of Crowding In India

India is overwhelmingly crowded and high density living conditions in majority of cases is prolonged, acute and inescapable. On the eve of World Population Day a leading newspaper of India *Hindustan Times* (July 11th, 2003) has reported, India has over 1.06 billion people, more than all Africa put together. Every sixth person on earth is an Indian. We are adding 33 babies a minute. Granted the fact that India is overpopulated, there has clearly been a paucity of research on crowding which is a direct resultant of population growth and density and related aspects in the Indian context. Jain (1976) is probably the first Indian scholar to have initiated systematic studies on crowding. His series of experimental and correlational studies are published in two volumes (Jain, 1987; Jain, 1997). In his initial study on the psychological consequences of crowding (Jain, 1976) selected competition tolerance as a cultural specific variable and examined its linkages with parameters of high density and need intensity. It was argued that in the Indian context scarcity of resources increase the competition tolerance which influences the individual's adaptability to high density conditions. Jain (1987) conducted an experimental study to examine the effects of resources on the feeling of crowding and personal space.

Manipulating high and low social density and adequate and scarce resources, the subject was tested for the feeling of crowding and selection of interpersonal distance with regard to strangers. Results indicated that under the condition of resource scarcity subjects exhibited greater feeling of crowding relative to their counterparts who were in the adequate resource conditions. The researcher suggested that under scarce resource condition the effects of high density are more pronounced as compared to adequate resource conditions. Even for day to day needs increased competition under scarce resource and high density condition motivates the respondents to develop more competition tolerance (Jain, 1978). Both high density and low socio-economic status failed to show any adverse effect on measures pertaining to expected helping and interpersonal attraction (Jain, 1987). Keeping the Indian societal characteristic including values and norms, collectivistic *versus* individualistic self, into consideration the results obtained by Jain (1987) seems meaningful largely because such values aid people to deal with crowding. The socio-cultural perspective should be taken into account to explain more comprehensively the crowding related effects.

In his subsequent study Jain and Surajmal (1983) reported that high density was related to crowding experience and has adverse affective consequences. The effects of high density were found to vary with respect to the participant's socio-economic status. The subjects belonging to low socio-economic status exhibited greater helping tendency and high feeling of crowding than their counterparts belonging to high socio-economic categories. Finally, evidence of ill health was found to be more in high density relative to low density conditions. Consistent results were obtained in other studies where high density was found to be associated with adverse emotional and health related problems (Preet Kamal & Jain, 1984; Nagar, 1985). The study of Tripathi (1986) conducted in Varanasi reported that high density was found to be associated with maladjustment, anxiety and neuroticism. In Ahmedabad, Kulkarni (1984) reported evidence of high crime rate in the areas where density level was high. In a similar vein, home crowding in Gorakhpur was found to be associate with various adjustment related problems (Nagar, 1985; Pandey & Nagar, 1988). For instance, Nagar (1985) attempted to examine the influence of overcrowding in rural/urban homes on interpersonal climate, adjustment and emotional states. The results of this study indicate that prolonged exposure to dwelling crowding was aversive and led to various adjustment related problems. The results further indicate that urban respondents as compared to their rural counterparts perceive their dwelling climate as more supportive (e.g. more involvement, more independence, more opportunity of intellectual growth) and reported less problems of adjustment. Finally, the results indicate that the adverse affect of crowding is due to the non-supportive dwelling climate. It seems that dwelling climate affects the relationship between crowding and problems of adjustment and that supportive home climate can possibly reduce the negative effects of crowding.

Two interrelated experimental studies on crowding were conducted by Nagar (1985). In the first study crowding was equated with density while in the second study both objective density and subjective experience of crowding were taken together and their relationship with task performance and affect was observed (Nagar & Pandey, 1987; Nagar, Pandey; & Paulus, 1987). The major results of these studies revealed that performance of cognitively complex tasks were adversely affected in high density conditions relative to low density conditions. However, variation in density failed to affect the performance of simple task. The results obtained in the second study revealed that high experience to crowding subjects performed significantly better on complex tasks relative to their low experience to crowding counterparts. This study also reported cross-situational effect in which the subjects who had high experience of crowding in their residential environment performed better on the cognitive tasks and exhibited more positive affect in high density laboratory conditions as compared to their counterparts who were from the low experience to crowding group. Other researchers have also reported that generalization is seen not only across situations but also across different type of environmental stressors having common elements (Lepore, Evans, & Palsane, 1991). Finally, as compared to objective crowding, subjective experience of crowding was found to be more potent predictor of outcome variables related to emotional states and health (Nagar, 1999; Jain, 1997; Nagar & Mishra, 2001).

Crowding research in the initial stages were primarily focused on objective indicators of crowding and the subjective judgments were not given due importance by the Indian researchers (Nagar *et al.*, 1988). It was felt that that a comprehensive understanding of various facets of experience of crowding goes beyond the simple density measures and taps more complex cognitive perceptual features of the dense environmental settings. Based on the contemporary theoretical models of crowding, attempt was made to construct the crowding experience scale. Both exploratory and confirmatory factor analyses were performed on a large data sets, which reflected evidence of four distinct components of crowding experience. These four components of crowding experience were labelled as—space satisfaction, positive or supportive relationship, negative or disruptive relationship and uncontrolled disturbance. These four components uniquely predicted outcome measures including psychological well-being, affect and symptoms (Nagar & Paulus, 1997). For instance, the uncontrolled disturbance was the most potent predictor of reported symptoms, with negative relationship further adding to its predictive power. Space satisfaction and negative relationship in the home both predicted psychological well-being and emotional state. To a large extent, similar components of crowding experience are reported by Jain (1997). For instance, the study of Jain (1997) reported evidence of three components of feeling of crowding, labelled as negative affect, arousal and loss of control.

A number of studies have examined contextual effects in response to high density situations. Tripathi (1990) in his study selected a number of day-to-day crowded situations in various laboratory, imaginary and life settings. Their results indicated that the feeling of crowding is not only situational specific but also depends on the sequential mode of presentation of setting. The interpersonal climate in the home was related to various types of adjustment behaviours. It seems that interpersonal climate in the home provided a context to reduce the adverse affective states of the respondents (Nagar, 1998). In other words it seems that the home climate mediates between density and adjustment. Some studies have done systematic empirical inquiry about the mediator/moderator variables in the residential settings (Evans, Palsane, Lepore, & Martin, 1989). In an interesting study conducted by Ruback and Pandey (1992) on autorickshaw travellers in India, the effects of temperature and crowding was investigated. They interviewed passengers under three levels of temperature and three levels of crowding and found significant main and interaction effects on effective measure. Higher temperature and more crowding produced more negative effect. They then manipulated the effects by telling some passengers about the effects of temperature, some about the effects of crowding and some were not told anything. The results revealed that knowing about the effects of temperature produced a greater sense of control and mediated the negative affect. However, being aware about the effects of crowding does not enhance the respondent's ability to cope with the adverse affect. This is supportive of the role of appraisal of cause effects in the stress process (Halpern, 1995). Jain and Palsane (2004) in their review article have indicated that any variable can be exclusively considered as mediator or moderator largely because the human-environment relationship is subjected to change along with the interaction. Their contention receives support in some studies (Lepore *et al.*, 1990; Lepore *et al.*, 1991). Lepore *et al.* (1991) compared social hassles and crowding in the home, in American and Indian samples, and found that although there was a relationship between the two, crowding was a better predictor of psychological distress.

A large-scale survey was carried out in three villages near Allahabad by Ruback and Pandey (1993). The results of this study revealed consistent gender differences with males relative to females exhibited more perceived control, more internal control, fewer physical symptoms and positive home evaluation. Interestingly, men reported more negative reactions to crowding. Women were less upset by the high social density and had desire for more people and children in their homes. Looking at the results it seems that regardless of the psychological costs, village culture demands women to have concern with social relationships and preference for male child. More research is needed on rural population to understand the role of cultural values and life styles in relation to crowding and its consequences. Scanning the literature it seems that Indian researchers have not paid adequate attention to culture specific coping styles to deal with crowding. Whatever, little work is done in more or less based on studies done in Euro-American context. However, in Indian setting some attempts were made. For instance, Padmavati (1991) investigated the effects of density on coping styles and reported no effects of density variation across differential coping styles. Through her study it was revealed that that both emotional and problem focused styles were equally preferred by the respondents residing in high and low density conditions. Jain and Palsane (2004) recommends that culture specific coping behaviours including social sharing, day-to-day quarrels with neighbours and family members should be examined to understand and explain the crowding related effects.

The overview of major research studies in the area of crowding is impressive. However, it is essential to focus on studies in range of settings like slums, rural areas, and low income housing, multistoried complexes. The socio-cultural realities, values and traditions may be considered to explain the effects of crowding and may be helpful in explaining how people deal with crowded situations of varied nature in a range of settings.

Theoretical Perspectives on Crowding

Several models and theoretical perspectives, discussed by researchers are available in edited books in the form of review articles (Baum & Paulus, 1987; Paulus & Nagar, 1989). In this section attempt is been made to discuss four major theoretical perspectives so as to provide an understanding of antecedents, psychological experiences and behavioural manifestations of crowding. The major theories of crowding are discussed are given below:

1. Overload theory
2. Behavioural constraint theory
3. Ecological orientations
4. Social interaction demand model

Overload Theory

Stimulus overload analyses of crowding have been based on earlier sociological theories of urban life developed by Wirth (1938). According to Wirth (1938) the urban areas can be characterized in terms of their size, density and the heterogeneity of the encompassed populations. These characteristics often result in the exposure of city residents to excessive levels of physical and social stimulation. The surfeit of stimulation, resulting from congestion and noise, is linked to a number of consequences at the community level—a cultural lifestyle of impatience, aloofness, social isolation and elevated levels of physiological and psychological ill effects.

An extension of earlier sociological theory was made by Milgram (1970). He defined overload as a situation in which the amount and rate of environmental inputs impinging on an organism exceeds the capacity to cope with them. He believes that individuals must enact certain insulative strategies under the conditions of overload, if he is to operate effectively in the urban environment. The strategies of adaptation to overload are disregard to low priority inputs, allocation of less time to each input, and development of aloof orientation towards strangers. Thus, an individual spends less time with each social contact and develops more superficial interpersonal relationships with others.

Application of overload model to the issue of crowding is reflected in the work of many psychologists (Desor, 1972;

Zlutnick & Altman, 1972). Altman (1975) has characterized crowding as a condition in which interpersonal control boundary mechanisms break down, such that achieved privacy is less than the level of privacy desired. A number of scholars have examined the overload interpretation of crowding. Desor (1972) utilizing a role-playing technique asked his subjects to place miniature figurines in a model room until they felt that the room was just short of being crowded. The number of entrance to the room and the presence and absence of partitions were varied. She found that more figurines were placed in rooms where there were partitions present and fewer external entrances. This result is consistent with the Altman's model, since partitions and entrances regulate social contact. Similarly, Valins and Baum (1973) found that students tended to rate corridor-style dormitories more crowded than suit-style dorms because the former design is less efficient in shielding residents from unwanted social encounters and stimulations.

Consistent with the assumption that high density situations are overloaded with complex information which may impair environmental perceptions. Saegert (1978) reported that Manhattan department store customers shopping for shoes were less able to recall details concerning to merchandise and layout of the shoe department under high density relative to low density conditions. Thus, one can infer that perceptual and behavioural deficits will occur in high density situations where individuals were unable to avoid excessive environmental stimulation. In another study Cohen (1978) presented a model of attention overload. His basic hypothesis states that in stress provoking environments, demands on attention capacity are created by the intensity, unpredictability and uncontrollability of the stressor. This increased demand might occur because individuals are required to continually monitor potentially threatening stimuli to evaluate their adaptive significance and to decide on appropriate coping responses. Cohen pointed out that the presence of others, whether anticipated or actually perceived, places high levels of demand on subject's attention and coordination. He further states that capacity of attention is not fixed but shrinks when there are prolonged demands.

This shrinkage presumably increases with both attention overload of an activity and duration of the activity.

Thus, in short the overload is concerned with interference and disruption of perceptual or cognitive information processing and when in high density condition the individuals are unable to avoid excessive stimulation, perceptual and behavioural disorders are seen.

Behavioral Constraint Theory or Social Interference Model

The behavioural constraint explanation on human crowding derived largely from Brehm's theory of psychological reactance and Sommer's conceptualization of personal space. Brehm (1966) states that maintenance of freedom of choice is an important motivating factor in human behaviour and perception. He argued that people are disposed to maintain or restore freedom when it is threatened and that individual's reaction to a setting is dependent on his success at accomplishing this goal. Sommer's (1969) notion of personal space is related to the construct of reactance. Personal space is conceived as an invisible "emotionally charged" zone surrounding each person, into which intruders may not come. Personal space presupposes the existence of individual's needs and situational norms, which establishes the bounds of comfortable interactional distance. Violation of spatial needs and norms create within the individual a reactance like pattern of emotional distress and the person will adopt certain ways of adjustment in order to preserve his personal space. The theory of reactance has received empirical support and many scholars who have explained their results of crowding within the social interference framework (Stokols, *et al.*, 1973; Proshansky *et al.*, 1970).

An important assumption underlying behavioural constraint analysis of crowding is that mere anticipation of social interference from others in one's immediate area plays an important role in determining the psychological and behavioural impact of high density. Moreover, it is assumed that the anticipation of crowding is mediated by social, personal and spatial variables. The findings of many studies are in consonance with this aforesaid assumption (Worchel & Teddlie, 1976).

The effects of social structural factors on perceived crowding and spatial needs are reflected in a study conducted by Stokols *et al.* (1973) in which they observed that the perception of crowding was greater when subjects played a game under competitive conditions, then when they adopted a cooperative set towards each other. Competitive task was viewed as a source of social interference in that it necessitates the monitoring of other behaviour whose interests conflict with ones own, and thereby, makes the need to "keep others at a distance" more salient.

Sundstrom (1975) manipulated room size, goal size and goal blocking independently in a experimental setting. Subjects reported greater irritation when goals are blocked then when they were not. It was noted that high density condition may cause disruption of interpersonal interactions, which subsequently produces psychological stress. Density is related to crowding when it is expected to reduce the individual's control over the environment. Crowding does not occur when it does not affect behaviour or goal blocking.

Ecological Orientations Model

In the late nineteenth century, ecological perspective was articulated by biologists and subsequently by sociologists. Barker (1960) applied the ecological theory to understand human social behaviour. He proposed two ecological theories—undermanning and overmanning theory. In the undermanning setting, there are insufficient personnel to carry out essential tasks and thus the available personnel are pressurized for more intensive participation. Thus, in the undermanning settings there are fewer participants than the number of available roles. Pressure induces the members to take on a variety of tasks and work longer hours. The predictions of undermanning theory have received support from a variety of studies involving comparison of large and small schools (Barker & Gump, 1964), churches (Wicker, 1969).

Barker (1960) theory of undermanning was extended by Wicker, McGarth and Armstrong (1973) through a construct overmanning. Overmanning setting was defined as those in which the number of eligible participants exceeds the personal capacity of the system. In the overmanning settings individuals perceive themselves as less valuable to the group and the setting. Thus, overmanning theory focuses on the limited availability of social roles within a behavioural setting.

Social Interaction Demand Model

The model proposed by Cox, Paulus, and McCain (1984) states that the major factors responsible for crowding effects lie within the dynamic properties of social interactions. The experience to crowding and concomitant behavioural and psychological effects are due in large part to the uncertainty, goal interference and cognitive overload experienced in high density setting. Furthermore, the model reveals that high density is not always stress provoking but may have positive effects when individual perceives no goal interference, or cognitive load or uncertainty in crowded conditions.

The studies summarized above suggest that often crowding is perceived as aversive and is related to performance deficits, negative mood states, ill health and reduced psychological well-being. However, the links between crowding and its consequences are dependent on range of factors like context, degree of control over the situation, prior exposure to crowding, social support, tolerance and so forth. Thus, a systematic study of mediators and moderators are needed to draw firm conclusions. Studies should also focus on those environmental settings where people are exposed to chronic crowding for a long period of time. For instance, in slums and low-income housing projects crowding coexists with other variables like poverty, education levels and so forth. These factors should be considered in designing studies on crowding. The unique coping styles, type of social support and tolerance are some factors that can help us in enhancing our understanding about crowding and its consequences.

Summary

Research on crowding has increased in voluminous proportion largely because in many cases its affects are adverse at the physical, social and psychological levels on both human and animal population. In this chapter an attempt is been made to present a brief historical overview of major social, psychological traditions and animal research that has dominated crowding research. Furthermore, while there is no all-inclusive definition of crowding so some popular conceptualization of crowding is presented so as to provide the reader major dimensions of crowding. In the next section a comprehensive review of major studies on crowding conducted on animals and human population are discussed. In this regard studies conducted in diverse settings ranging from home, neighbourhood to prisons utilizing both correlational and experimental studies are presented. The strength and limitations of these studies and major results are presented so as to derive some conclusion and possible generalizations.

A brief discussion is made to relate crowding and the degree to which people adapt to high density living conditions and how costly is the adaptation process. A number of studies are reviewed that examines the role of cultural factors and individual differences in response to crowded conditions. These studies have noted that not everyone is affected by crowding in the same way. The threshold of tolerance to crowding varies with respect to the degree of prior experiences of crowding at home or in crowded cities is noted by the researchers.

Granted the fact that the crowding literature probably represents one of the most extensive empirical literature in environmental psychology in the Indian context, thus a separate section is devoted to briefly review major studies of crowding in India. Finally, the major theoretical perspectives on crowding is presented. Although evidence exists for crowding effects in a broad range of environmental contexts, much remains to be learned about the impact of density on different types of populations (e.g. old aged homes, disadvantaged groups) and in different environmental contexts.

REFERENCES

Aiello, J.R., Baum, A., & Gormley, F. (1981). Social determinants of residential crowding stress. *Personality and Social Psychology Bulletin*, 7, 643-649.

Aiello, J.R., Epstein, Y., & Karlin, R. (1975). Effects of crowding on electrodermal activity. *Sociological Symposium*, 14, 42-57.

Aiello, J., Nicosia, G., & Thompson, D. (1979). Physiological, social and behavioral consequences of crowding on children and adolescents. *Child Development*, 50, 195-202.

Altman, I. (1975). *The Environment and Social Behavior*. Monterey: Brooks/ Cole.

Barker, R. (1960). *Ecology and Motivation*. Nebraska Symposium on Motivation.

Barker, R., & Gump, P. (1964). *Big School, Small School*. Stanford University Press.

Baron, R.M., Mandel, D.R., Adams, C.A., & Griffitt, L.M. (1976). Effects of social density in university residential environments. *Journal of Personality and Social Psychology*, 34, 434-446.

Baum, A., Aiello, J., & Calesnick, L. (1978). Crowding and personal control: Social density and the development of learned helplessness. *Journal of Personality and Social Psychology*, 36, 1000-1011.

Baum, A., Fleming, R., & Davidson, L. (1983). Natural disaster and technological catastrophe. *Environment and Behaviour*, 15, 333-354.

Baum, A., & Gatchel, R. (1980). *Cognitive Determinants of Reaction to Uncontrolled Events: Development of Reactance and Learned Helplessness*. Unpublished manuscript, Uniformed Services, University of Health Science.

Baum, A., & Greenberg, C.I. (1975). Waiting for a crowd: the behavioral and perceptual effects of anticipated crowding. *Journal of Personality and Social Psychology*, 32, 671-679.

Baum, A., Grunberg, N.E., & Singer, J.E. (1982).The use of psychological and neuroendocrinological measurement in the study of stress. *Health Psychology*, 1, 217-236.

Baum, A., & Koman, S. (1976). Differential response to anticipated crowding: Psychological effects on social and spatial density. *Journal of Personality and Social Psychology*, 34, 526-536.

Baum, A., & Paulus, P. (1987). Crowding. In D. Stokols & I. Altman (Eds.), *Handbook of Environmental Psychology* (Vol. 1, pp. 553-570), New York, John Wiley.

Baum, A., Singer, J.E., & Baum, C. (1982). Stress and the environment. In G.W. Evans (Ed), *Environmental Stress* (pp. 15-44), New York: Cambridge University Press.

Baum, A., & Valins, (1977). *Architecture and Social Behavior: Psychological Studies of Social Density*. Hillsdale, NJ: Erlbaum.

Baum A., & Valins, (1979). Architectural mediation of residential crowding and control: crowding and the regulation of social contact.

In L. Berkowitz (Ed.), *Advances in Experimental Psychology* (Vol. 12), pp. 131-175), New York: Academic Press.

Bergman, B.A. (1971). *The Effects of Group Size, Personal Space, and Success-Failure on Physiological Arousal, Test Performance, and Questionnaire Response*. Unpublished Doctoral dissertation, Temple University, Philadelphia.

Booth, A., & Johnson, D. (1975). *The Effects of Crowding on Child Health and Development*, Mimeo.

Booth, A., & Welch, S. (1973). *The Effects of Crowding: A Cross-National Study*. Unpublished manuscript, Ministry of State for Urban Affairs, Ottawa, Canada.

Brehm, J. (1966). *A Theory of Psychological Reactance*. New York: Academic Press.

Calhoun, J. (1962). Population density and social pathology. *Scientific American*, 206, 139-148.

Calhoun, J. (1973). Death squared: The explosive growth and demise of a mouse population. *Proceedings of the Royal Society of Medicine*, 66, 80-88.

Carson, R. (1962). *The Silent Spring*. Boston: Houghton Miffin.

Carstairs, G. (1969). Overcrowding and human aggression. In H.D. Graham & T.D. Curr (Eds.), *Violence in America: Staff Report to the National Commission on the Causes and Prevention of Violence*. New York: Signet Books.

Cassell, J. (1971). Human consequences of population density. *Rapid Population Growth : Consequences and Policy Implications*, Baltimore: John Hopkins Press.

Cassidy, T. (1997). *Environmental Psychology: Behavior and Experience in the Context*. Sussex, Psychology Press.

Chuaqui, C., Lemkau, P., Legarrefa, A., Contresas, M. (1966). *Suicide in Santiago, Chile*. Public health reports, 81, 1100-1117.

Christian, J. (1963). Endocrine adaptive mechanism and the physiologic regulation of population growth. In W. Mayor & R. Gelder (Eds.), *Physiological Mammalogy*, (Vol. 1) New York: Academic Press.

Christian, J., Flyger, V., & Davis, D. (1960). Factors in the mass mortality of a herd of silka deer nippen, *Chaspapeake Science*, 1, 19-95.

Cohen, S.A. (1978). Environmental load the allocation of attention. In A. Baum, J.E. Singer, & S. Valins (Eds.), *Advances in Environmental Psychology* (Vol. 1). Hillsdale: Erlbaum.

Commoner, B. (1963). *Science and Survival*. New York; Viking.

Cox, V., Paulus, P.B., McCain, G., & Schkade, W. (1979). Field research on the effects of crowding in prisons and on offshore drilling platform. In J. Aiello and A. Baum (Eds.), *Residential Crowding and Design*. New York: Plenum.

Cox, V., Paulus, P.B.,& McCain, G. (1984). Prison crowding research: The relevance of prison housing standards and a general approach regarding crowding phenomena. *American Psychologists*, 39, 1148-1160.

D'Atri, D. (1975). Psychophysiological responses to crowding. *Environment and Behavior*, 7, 237-252.

D'Atri, D., & Ostfeld, A.M. (1975). Crowding: Its effects on the elevation of blood pressure in a prison setting. *Preventive Medicine*, 4, 550-566.

Deevey, F. (1960). The hare and the haruspex. *American Scientist*, 48, 415-429.

Desor, J. (1972). Towards a psychological theory of crowding. *Journal of Personality and Social Psychology*, 11, 93-97.

Dooley, B. (1974). *Crowding Stress: The Effects of Social Density on Men with Close or Far Personal Space*. Unpublished doctoral dissertation, University of California, Los Angeles.

Ehrlich, P., Ehrlich, A. (1970). *Population Resources and Environment*. San Francisco: Freeman.

Epstein, Y., & Baum, A. (1978). Crowding: Methods of study. In A. Baum & Y. Epstein (Eds.), *Human Response to Crowding*, Hillsdale, NJ: Erlbaum.

Evans, G. (1975). *Behavioral and Physiological Consequences of Crowding in Humans*. Unpublished doctoral dissertation, University of Massachusetts.

Evans, G. (1979). Crowding and human performance. *Journal of Applied Social Psychology*, 9, 27-46.

Evans, G., & Lepore, S. (1992). Household crowding and social support: A quasi-experimental analysis. *Journal of Personality and Social Psychology*, 57, 994-999.

Evans, G., Palsane, M., Lepore, S., & Martin, J. (1989). Residential density and psychological health: The mediating effect of social support. *Journal of Personality and Social Psychology*, 57, 994-999.

Freedman, J. (1975). *Crowding and Behavior*. San Francisco: Freeman.

Freedman, J.L., Levy, A., Buchanan, R., & Price, J. (1972). Crowding and human aggressiveness. *Journal of Experimental Social Psychology*, 8, 528-548.

Freedman, J.L., Klevansky, S., & Ehrlich, P.I. (1971). The effect of crowding on human Task performance. *Journal of Applied Social Psychology*, 1, 7-26.

Fuller, T.D., Edwards, J.N., Vorakitphokatorn, S., & Sermsri, S. (1993). Household crowding and family relations in Bangkok. *Social Problems*, 40, 410-430.

Galle, O., Gove, W., & McPherson, J. (1972). Population density and pathology: What are the relationships for man? *Science*, 176, 2-30.

Gifford, R., Peacock, J. (1979). Crowding: More fearsome than crime provoking? Comparison of an Asian city and a North American city. *Psychologia*, 22, 79-83.

Glass, D., & Singer, J. (1972). *Urban Stress*. New York: Academic Press.

Goduka, I.N., Poole, D.A., & Aotaki-Phenice, L. (1992). A comparative study of black South African children from three different contexts. *Child Development*, 63, 509-525.

Goeckner, D., Greenough, W., & Meed, W. (1973) Deficits in learning tasks following overcrowding in rats. *Journal of Personality and Social Psychology*, 28, 256-261.

Gove, W., & Hughes, M. (1983). *Overcrowding in the Households*. New York: Academic Press.

Gove, W., Hudges, M., & Galle, O. (1979). Overcrowding in the home: An empirical investigation of its possible pathological consequences. *American Sociological Review*, 44, 59-80.

Halpern, D. (1995). *Mental Health and the Built Environment*. London: Taylor & Francis.

Hauser, P., & Dunken, O. (1959). *The Study of Population: An Inventory of Appraisal. Chicago*: University of Chicago Press.

Hutt, C., & Vaizey, M.J. (1966). Differential effects of group density on social behavior. *Nature*, 209, 1371-1372.

Jain, U. (1976). *Competition Tolerance and Need Hierarchy Patterns as Related to Population Growth*. Project Report of the ICSSR.

Jain, U. (1978). Competition tolerance in high-low density in urban-rural areas. *Journal of Social Psychology*, 105, 297-298.

Jain, U. (1987). *The Psychological Consequences of Crowding*. New Delhi: Sage.

Jain, U. (1997). *The feeling of crowding and its psychological consequences*. New Delhi: Har-Anand Publications.

Jain, U., & Palsane, M.N. (2004). ICSSR survey in Psychology, Vol. 3, New Delhi: Sage.

Jain, U., & Suraj, M. (1983). Effects of prolonged deprivation on attribution of causes of success and failure. *Journal of Social Psychology*, 124, 143-49.

Jacobs, J. (1961). *The Death and Life of Great American Cities*. New York: Random House.

Johnson, D., & Matross, R. (1977). Interpersonal influence in psychotherapy: A Socio-Psychological view. In A. Gurman & A. Razin (Eds.), *Effective psychotherapy: A Handbook of Research*. New York: Pergmon.

Jorgenson, D.O., & Dukes, F.O. (1976). Deindividuation as a function of density and Group membership. *Journal of Personality and Social Psychology*, 34, 24-39.

Karlin, R., Epstein, Y., & Aiello, J. (1978). *Strategies for investigation of crowding*. In A. Esser., & B. Greenberg (Eds.), designs for community and privacy, New York: Plenum Press.

Klein, K., & Harris, B. (1979). Disruptive effects of disconfirmed expectancies about crowding. *Journal of Personality and Social psychology*, 37, 391-395.

Korte, C. (1979). Helpfulness in urban environment. In A. Baum., J.E. Singer, & S. Valins (Eds.), *Advances in Environmental Psychology: The Urban Environment* (Vol 1). Hillsdale, NJ: Erlbaum.

Kulkarni, K. (1984). *Geography of Crowdıng and Human Response*. New Delhi: Concept.

Lawrence, (1974). Science and sentiment: Overview of research on crowding and human Behavior. *Psychological Bulletin*, 712-720.

Layhausen, P. (1965). The same community: A density problem? *Discovery*, 26, 27-33.

Lepore, S., Evans, G., & Schneider, M. (1991). Dynamic role of social support in the link Between chronic stress and psychological distress. *Journal of Personality and Social Psychology*, 61, 899-909.

Lepore, D., Merritt, k., Kawasake, N., & Mancuso, R. (1990). *Social Withdrawal in Crowded Residences*. Paper presented at the meeting of the Western Psychological Association, Los Angeles, CA.

Lestor, D. (1970). Social disorganization and completed suicide. *Social Psychiatry*, 5, 175-176.

Levy, L., & Herzog, A. (1974). Effects of population density and crowding on health and social adaptation in the Netherlands. *Journal of Health and Social Behavior*, 15, 240.

Loo, C. (1973). Important issues in researching the effects of crowding on humans. *Representative Research in Psychology*, 26, 27-33.

Loring, W. (1955). Housing characteristics and social disorganization. *Social Problems*, 3, 160-168.

Lundberg, U. (1976). Urban commuting: Crowdedness and catecholamine excretion. *Journal of Human Stress*, 2, 26-32.

MacDonald, W., & Oden, C. (1973). *Effects of Extreme Crowding on the Performance of Five Married Couples During Twelve Weeks of Intensive Training*. Proceedings of the 81st Annual Convention of the American Psychological Association, 8, 209-210.

Mardson, H. (1972). Crowding and population density. In J.F. Wohlwill & D. H. Carson (Eds), *Environment and the Social Sciences: Perspectives and Applications*, APA, Washington, D.C.

Marsella, A., Escudero, M., & Gorden, P. (1970). The effects of dwelling density on mental disorders in Felipino men. *Journal of Personality and Social Psychology*, 11, 288-294.

McCain, G., Cox, V., & Paulus, P.B. (1976). The relationship between illnesses complains and degree of crowding in a prison environment. *Environment and Behavior*, 8, 289-291.

McCarthy, D., & Saegert, D. (1978). Residential density, social overload, and social withdrawal. *Human Ecology*, 6, 253-272.

McCauley, C., Coleman, G., & DeFusco, P. (1977). *Commuters' Eye Contact with* Strangers in City and Suburban Train Stations. Evidence of Short-Term Adaptation to *Interpersonal Overload in the City*. Unpublished manuscript, Bryn Mawr college, Bryn Mawr, PA.

McCauley, C., & Newman, J. (1977). Eye contact with strangers in a city, suburb, and small town. *Environment and Behaviour*, 9, 547-558.

Milgram, S. (1970). *The experience of living in big cities. Science*, 167, 1461-1468.

Mitchell, R. (1971). Some implications of high density housing. *American Sociological Review*, 36, 1461-1468.

Nagar, D. (1985). *Experiences and consequences of crowding*. Unpublished doctoral dissertation, Allahabad University, Allahabad.

Nagar, D. (1998). *Human Reactions to Crowding*. Jaipur: Printwell Publishers.

Nagar, D. (1999). Environmental stress and human behavior. In G. Misra (Ed), *Stress and Health (pp. 238-262)*, New Delhi: Concept.

Nagar, D., & Mishra, M. (2001). Slum rehabilitation in Bhopal: Its psychological Consequences. *Indian Cooperative Review*, 38, 171-183.

Nagar, D., & Pandey, J. (1987). Affect and performance of cognitive task as a function of Crowding and noise. *Journal of Applied Social Psychology*, 17, 147-157.

Nagar, D., Pandey, J., & Paulus, P. (1988).The effects of residential crowding experience On reactivity to laboratory crowding and noise. *Journal of Applied Social Psychology*, 18, 1423-1442.

Nagar, D., & Paulus, P.B. (1997). Residential crowding experience scale: Assessment and validation. *Community and Applied Social Psychology*, 7, 303-319.

Odera, P., & Hasan, Q. (1993). Assessment of difficulties of foreign students in India. *Journal of Personality and Clinical Studies*, 9, 25-30.

Padmavati (1991). *Coping with Crowding Stress*. Unpublished doctoral dissertation, Barkatullah University, Bhopal.

Pandey, J. (1978). Effects of crowding on human social behavior. *Journal of Social and Economic Studies*, 6, 85-95.

Pandey, J., & Nagar, D. (1987). Experiences of indwelling and outdwelling crowding and social psychological consequences. *Social Science International*, 3, 1-9.

Pandey, J., & Nagar, (1988). Residential crowding in India: Its social psychological consequences. Paper presented at the 24th International Congress of Psychology, Sydney.

Paulus, P.B. (1988). *Prison crowding: A Psychological Perspective*. New York: Springer-Verlag.

Paulus, P.B., Annis, A. B., Seta, J.J., Schkade, J.K., & Matthews, R.W. (1976). Crowding does affect task performance. *Journal of Personality and Social Psychology*, 34, 248-253.

Paulus, P.B., McCain, G., & Cox, V. (1978). Death rates, psychiatric commitments, blood pressure, and perceived crowding as a function of institutional crowding. *Environment and Behavior*, 3, 107-116.

Paulus, P.B., McCain, G., & Cox, V. (1981). Crowding in jails. *Basic and Applied Social Psychology*, 4, 89-107.

Paulus, P.B., & Nagar, D. (1989). Environmental influences in groups. In P.B. Paulus (Ed.), *Psychology of Group Influences*, Hillsdale, NY; Erlbaum.

Powell, L. (1990). Factors associated with the under-representation of African-American In mathematics and science. *Journal of Negro Education*, 59, 292-298.

Preet, K., & Jain, U. (1984). *Effects of Population Density on Mental Health.* Paper presented at Indian Psychiatric Conference, Udaipur.

Proshansky, H., Ittelson, W., & Rivlin, L. (1970). Freedom of choice and behavior in a physical setting. In H.M. Proshansky *et al.* (Eds.), *Environmental Psychology: Man and his Physical Setting,* New York: Holt, Rinehard & Winston.

Reddy, D., Baum, A., Fleming, R., & Aiello, J. (1981). Mediation of social density by Coalition formation. *Journal of Applied Social Psychology,* 11, 529-537.

Rodin, J. (1976). Crowding, perceived control, and response to controllable and uncontrollable outcomes. *Journal of Experimental Social Psychology,* 12, 133-158.

Rohe, W., & Patterson, A.H. (1974). *The Effects of Varied Levels of Resources and Density on Behavior in a Day Care Center.* Paper presented by the Environmental Design Research Association, Milwaukee, WI.

Ruback, R.B., & Pandey, J. (1991). Crowding, perceived control, and relative power: An analysis of households in India. *Journal of Applied Social Psychology,* 21, 345-374.

Ruback, R.B., & Pandey, J. (1992). Very hot and really crowded: Quasi-experimental investigations of Indian "Tempos". *Environment and Behavior,* 24, 527-554.

Ruback, R.B., & Pandey, J. (1993). *Gender Differences in Reaction to Household Crowding: Affiliation and Stress in Rural India.* Paper presented at the Annual convention of National Academy of Psychology, Delhi University, Delhi.

Saegert, S. (1978). High density environments: their personal and social consequences. In A. Baum & Y. Epstein (Eds), *Human Responses to Crowding,* (pp. 257-281), Hillsdale, NJ: Erlbaum.

Saegert, S. (1974). *Effects of Spatial and Social Density on Arousal, Mood and Orientation.* Unpublished doctoral dissertation, University of Michigan, Ann Arbor.

Sainsbury, P. (1955). *Suicide in London.* London: Chapman and Hall.

Schmidt, D., & Keating, J. (1979). Human crowding and personal control: An integration of the research. *Psychological Bulletin,* 86, 680-700.

Schmitt, R. (1963). Implications of density in Hong Kong. *Journal of American Institute of Planners,* 29, 210-217.

Schmitt, R. (1966). Density, health and social disorganization. *Journal of American Institute of Planners,* 32, 38-40.

Schwab, J., Nadeau, S., & Warheit, G. (1979). Crowding and mental health. *Pavlovian Journal of Biological Science,* 14, 226-233.

Seligman, M.E. (1975). *Helplessness.* San Francisco: Freeman.

Sherrod, D.R. (1974). Crowding, perceived control and behavioral aftereffects. *Journal of Applied Social Psychology,* 4, 171-186.

Singer, J.E., Lundberg, U., & Frankenhaeuser, M. (1978). Stress on the train: A study of Urban commuting. In A. Baum, J.E. Singer, S. Valins (Eds.), *Advances in Environmental Psychology* (Vol. 1). Hillsdale: NJ: Erlbaum.

Sinha, S.P., Nayyar, P., & Mukherjee, N. (1995). Perception of crowding among children and adolescents. *Journal of Social Psychology*, 135, 263-268.

Sommer, R. (1969). *Personal Space: The Behavioral Basis of Design.* Englewood Cliffs, NJ: Prentice-Hall.

Southwick, C.H. (1967). An experimental study of intra-group antagonistic behavior in Rhesus monkeys. *Behavior*, 28, 182-209.

Stokols, D. (1976). The experience of crowding in primary and secondary environment. *Environment and Behavior*, 1, 49-86.

Stokols, D., & Ohlig, W. (1975). *The Experience of Crowding under Different Social Climates.* Paper presented at the meeting of the American Psychological Association, Chicago.

Stokols, D., Rall, M., Pinner, B., & Schopler, J. (1973). Physical, social, and personal determinants of the perception of crowding. *Environment and Behavior*, 5, 87-115.

Sundstrom, E. (1975). The experimental study of crowding: Effects of room size, intrusion and goal blocking on non-verbal behaviors, self-disclosure, and self-reported stress. *Journal of Personality and Social Psychology*, 32, 645-654.

Thompson, W. (1953). *Population Problem* (fourth edition). New York: McGraw-Hill.

Tripathi, S. (1986). *Effects of Crowding on Human Behavior.* Unpublished Doctoral Dissertation, Kashi Vidhyapeeth, Varanasi.

Tripathi, H. (1990). *Cognitive and Affective Consequences of Crowding Stress.* Unpublished Doctoral Dissertation, Bhopal university, Bhopal.

Valins, S., & Baum, A. (1973). Residential group size, social interaction, and crowding. *Environment and Behavior*, 5, 421-439.

Webb, W. M., & Worchel, S. (1993). Prior experience and expectation in the context of Crowding. *Journal of Personality and Social Psychology*, 65, 512-521.

Wicker, A. (1969). Size of church membership and members support of church behavior settings. *Journal of Personality and Social Psychology*, 13, 278-288.

Widmayer, S.M., Petersen, L.M., Larner, M., & Carnnhan, S. (1990). Predictors of Haitian-American infant development at twelve months. *Child Development*, 61, 410-415.

Winsborough, H. (1965). The social consequences of high population density. *Law and Contemporary Problems*, 30, 120-126.

Wirth, L. (1938). Urbanism as a way of life. *The American Journal of Sociology*, 44, 1-24.

Worchel, S., & Teddlie, C. (1976). The experience of crowding: A two factor theory. *Journal of Personality and Social Psychology*, 34, 30-40.

Zlutnick, S., & Altman, I. (1972). Crowding and human behavior. In J. Wohlwill & D. Carson (Eds.), *Environment and the Social Sciences*, Washington, D.C. American Psychological Association.

6

ENVIRONMENTAL STRESS AND HEALTH

The issues pertaining to environmental stress have assumed a critical dimension in the technocratic industrial societies. Very often reports dealing with the adverse effects resulting from change in the environmental balance, global earth warming, acid rain, acidification of ground, chemical pollution, noise, crowding are heard. The adverse effects of these ecological irritants have heightened the awareness of community-environment crisis and have perplexed the scientists, administrators, policy makers, and social workers to think and take necessary steps towards resolving the problems of ecological balance. Although, rapid industrialization, which is one of the prime sources of man-environment disharmony, has assured a number of economic benefits, however, danger prevails. Even the preliminary safeguards to ensure safety and health are lacking in many of the dangerous chemical plants. Our past experiences conform the fact that zero-order risk free chemical and nuclear industries are inconceivable. Sufficient evidence exists that not only the people of the third world countries were victimized by the industrial accidents but also countless marine and human casualties were reported in the most advanced countries of the world like Soviet Union, United States and Germany in the 1980s.The suddenness and intensity by which the global eco-catastrophe adversely affects the big strata of the community is immediately noticed not only because of the incalculable human and material cost but also due to greater publicity. However, people fail to notice or are unaware of the dangerous psychological effects of silent ecological stressors, which impinge on them in their day-to-day lives. In this chapter attempt is been made to examine the impact of major environmental stressors (e.g. noise, air and chemical pollution) which pose potential threats to cognitive, emotional, behavioural and health related aspects on humans.

In recent years the area of psychological stress has made significant advances, in understanding the relationship between physical environmental influences on health and well being. It deals with the impact of a wide range of mostly negative outcomes that otherwise seem to defy adequate explanation. Advances are been made in understanding the intricacies of environmental stress with the influential work of Selye (1956) and Lazarus (1966). Furthermore, a series of well designed studies of Glass and Singer (1972) on noise and studies on chemical pollution by Veitch and Arkkelin (1995); the environmental stress concept has led to ample amount of research on the properties of environment and its relationship with human behaviour and health. Another prominent concerns of environmental psychology for the last two decades has been the understanding of underlying psychological processes through which physical environment and behaviour are linked. A number of studies have noted the moderating/ mediating role of various psychological factors like perceived control, stress tolerance, coping resources and personality dispositions of environmental factors on human social behaviour, health and well-being (Baum & Paulus, 1987; Evans & Cohen, 1987; Nagar, 1999; Jain and Palsane, 2004).

In this chapter focus will be on the physical environmental conditions that we experience in our daily life. The chapter is organized into three sections. A brief overview of stress paradigm including the definition of stress, characteristics of environmental stressors and general theoretical perspectives in stress literature will be presented. The next section briefly summarizes major type of environmental stressors. In this section attempt will be made to review studies on noise, air and chemical pollution. Finally, in the last section the status of environmental stress research will be evaluated.

Overview of Stress Paradigm

It is very difficult to define the concept of stress. The convention definitions of stress vary depending on either situational factors interfering with the desired goals or emphasizing the response to stressors. Both these approaches to understanding environmental stressors suffer from several methodological problems (Evans & Cohen, 1987; Lazarus & Cohen, 1977). In the past process oriented models of stress (Lazarus & Folkman, 1984) have emerged which take into account the psychological reality of the stress processes more adequately. Such an approach view stress as a dynamic transaction between the person and his environment mediated by several variables like cognitive appraisal and coping. According to Lazarus and Folkman (1984) the mediating constructs (e.g. appraisal, coping) must be studied as processes, occurring over time that continuously alter or change the relationship between environmental stressors and related outcomes. The process-oriented models tend to examine how the individual appraises the misfit between environmental attributes and human needs. The severity of the stress and its long-term consequences for the individual are affected by many factors, like characteristics of the stressful situation, importance of the misfit to the individual, perceived awareness of the misfit, and the coping strategies available to the individual for establishing congruence between individual's need and environmental demands (Evans & Cohen, 1987). Before discussing the sources of environmental stressors it is pertinent to examine the contemporary perspectives and psychological mediation of environmental stressors.

Physiological Perspective

The first systematic treatment of stress phenomenon was developed by Selye (1956). The General Adaptation Syndrome (GAS) on which his physiological model is based includes three stages: alarm, resistance and exhaustion. His model focuses on bodily defence against physical challenge to maintain a state of homeostasis.

When first exposed to a stressor, the body responds by mobilizing its coping abilities, the alarm reaction represents preparation for resistance. During this phase the endocrine system secretes adrenaline and neoadrenaline into the blood stream, which stimulates the adrenal cortex to produce various chemical substances including the secretion of anti-inflammatory corticosteroids. When the adaptive reserves are made ready, the body enters a stage of resistance to combat threats posed by the invading pathogen by setting up a feedback loop by stimulating the adrenal medulla to release more catecholamines. As the pathogen is invaded the adrenal medulla returns to normal level and adaptation is achieved. When the threats are repeated frequently after short intervals, the organism remains in a state of arousal until adaptation is attained or the adaptive reserves are depleted bringing the organism into a stage of exhaustion. At this stage psychological dysfunction occurs and the body suffers breakdown or disease.

Selye (1956) in his model has made no attempt to distinguish between responses to physiological and psychological stressors and has not analyzed the mechanism that triggers the GAS. In contrast, Mason (1968) has argued the relationship between psychological response and emotional arousal as a specific stress response. He has also suggested that psychological threats are necessary for adrenal stimulation in the stress response. Even relatively minor changes in the environment are sufficient in activating the hypothalamus for adrenal secretion. Frankenhaeuser and Johansson (1986) has discussed extensively the role of adrenal system explaining the causes of psychological arousal, effort and distress in work environment under the condition of stimulus underload as well as stimulus overload. Finally, evidence exists regarding the coping processes, which may influence the psychological responses to stress.

Psychological Perspectives

The importance of psychological variables both as mediators of physical stressors (Glass & Singer, 1972) and as independent stressors (Nagar, 1999) has been increasingly emphasized by researchers. Responses to stressors are evaluated by the degree to which they are perceived as potentially harmful or threatening. The primary appraisal of each stressor will depend upon a number of personal and situational factors including attitudes towards the stressor, evaluation of its costs and rewards, knowledge about its consequences, prior experience about the aversive properties of the stressor, the duration of the stressor and the degree of controllability. If the individual appraises the stressful situation as upsetting, threatening and harmful, the secondary appraisals are made. During these processes people turn their attention to different modes of coping with the threat that may be either manipulative or accommodative.

Individuals may take direct action by trying to alter or manipulate his/her relationship to the stressful situation by removing the physical presence of the stressor to escape from the stressful event. Palliative coping becomes necessary when the person is not able to take direct action. Individual accommodates the stressful situation by changing his/her internal environment like creating psychological defence mechanism, taking drugs and so forth. These interactive coping modes are directed towards emotional regulation and may include reappraisal of a stressful situation. Coping may prove difficult when the stressful events are novel and ambiguous. However, when the stressors are familiar and efficacious coping strategies are readily available coping is likely to be successful.

Psychological Mediation of Environmental Stressors

Not everyone reacts similarly to the physical sources of stress. In fact people vary greatly in their reactions to environmental stressors like pollution and noise. The critical question is the accurate understanding of psychological mediators that intercede between objective measures of environmental stressors and the subjective or behavioural responses. While most of the scholars accept the mediating processes of appraisal. However, there is paucity of research examining the mediating variables in the context of noise and air pollution that may contribute towards explaining and predicting performance deficit, psychological well-being and health related outcomes.

Sources of Environmental Stressors

On the basis of eight major dimensions like perceptual salience, predictability, controllability, positive and negative value, suddenness and gradual onset, importance, type of adjustment required and duration of exposure, three major types of environmental stressors may be identified. These are: (1) Cataclysmic stressors, (2) Ambient stressors and (3) Daily hassles (Campbell, 1983; Evans & Cohen, 1987; Lazarus & Cohen, 1977).

Cataclysmic Stressors

Cataclysmic events refer to sudden catastrophes, unique and single powerful single event that demand major adaptive responses and affect large number of people. These types of events are broadly divided into two categories, war and disasters. These events include volcanic eruptions, earthquakes, floods and technological disasters. These disasters are all artificially created and are the result of human error, deliberate human action or technological breakdown or all the three.

Over the past two decades or so a series of disasters both natural and technological have claimed lives of thousands of people all across the world. For instance, the eco-catastrophe resulted from the leakage of toxic MIC gas at Bhopal is one example of cataclysmic environmental event. The occurrence of this environmental tragedy was sudden, powerful, uncontrollable and disastrous affecting the collective psyche of the entire Bhopal community in central India. In this disaster more than 3000 people died and over 2,00,000 others were injured in which thousands of the gas exposed victims are still suffering from serious mental, psychological and physical disorders (Weir, 1987; Hazarika, 1987; Everest, 1986; Jain & Palsane, 2004). A systematic and detailed account of Bhopal tragedy is available in The State of India's Environment Report. The report discusses mass panic, the plight of victims, health effects, failure of safety devices and failure of administration (Agarwal & Narain, 1985). Such cataclysmic events have consequences for a large section of society including survivors to the bereaved and those involved in provided rescue services.

In the aftermath of cataclysmic events attempts were made to identify the consequences for individuals. The expected consequences following the traumatic event ranged from shock, psychological disturbance and grief. Interestingly, two surprising results were also obtained. First, the results revealed that all psychological consequences were not always immediate, and often were present only after 9 to 12 months after the disaster. Secondly, the consequences were similar and as severe for workers involved at the scene as for the direct survivors who had actually been in the event. The psychological aftereffects of cataclysmic events are examined both in the east and in the west (Baum, Fleming & Davidson, 1983; Misra, 1988; ICMR, 1986).

Indian Council of Medical Research (ICMR) conducted an extensive survey on MIC gas victims immediately after the tragedy. The results of this survey revealed that MIC damaged the vital physiological systems including respiratory, ophthalmic, muscaloskeletal, neuro-psychaitric, and gastrointestinal. The survivors also suffered from Post-traumatic Stress disorders. A study conducted by Sethi, Sharma, Trivedi, and Singh (1987) found evidence of psychiatric morbidity in 22.6 per cent of gas exposed victims. The results of this study revealed anxiety neurosis, depressive neurosis, and adjustment reactions with prolonged deprivation in a large number of patients. After elapse of one year a community survey conducted by Tandon (1991) revealed prevalence rate of mental disorders upto 13.2 per cent. A longitudinal study over a period of three years was carried out by Misra (1988) on children belonging to three age groups: 4-6 yrs, 7-9 yrs and 10-12 yrs. The results of this study found evidence of mental disorders and reported adverse effects on intellectual, social and behavioural functions. Asthana and Jain (1986) reported that gas victims showed more alienation, powerlessness, hopelessness and self-estrangement relative to their counterparts who were from the non-exposed areas.

Ambient Stressors

The term ambient stressor refers to chronic and global conditions which are relatively more continuous and vary in intensity such as pollution, noise and crowding and temperature. The term 'ambient' connotes the background conditions. Ambient physical conditions generate psychological discomfort only when they block the attainment of goals. In these conditions ambient stressors come into the forefront of consciousness at cognitive, emotional and psychological levels. Another property of these stressors is that they are low on perceptual salience and their motivational salience is variable (Baum, Singer, & Baum, 1982; Evans, Jacobs & Frager, 1982).

Daily Hassles

Daily hassles range from the ordinary disturbance in the family and conflicts regarding work to various aspects of the physical environment (Caspi, Bolger and Eckenrode, 1987; Evans & Cohen, 1987). The accumulation of these irritants such as loosing your keys, dealing with colleagues, getting late to work can be much more stressful because of its cumulative effects. Daily hassles are characterized by more of less circumscribed events that are stable, non-urgent and appear to be for a short duration. Daily hassles have greater perceptual salience and variable motivational salience. Daily hassles may cause negative mood state, reduce well-being and health problems. For the most part, daily hassles are routine conditions and require only minor adaptive responses (Lazarus & Cohen, 1977). However, evidence exists that adaptive responses to daily hassles are not always minor in nature (Delongis, Coyne, Dakof, Folkman, and Lazarus, 1982).

Effects of Environmental Stressors

In this section of the chapter the effects of noise, air and chemical pollution on various outcome measures related to human performance, general well-being and health will be discussed.

Noise Pollution

In recent years noise is being considered as a very important stress factor. Very often, the term noise pollution signifies the vast array of sounds that are being produced, generating stress and creating health hazards. Noise is sometimes defined in terms of unwanted sound, in the wrong place, at the wrong time. People are exposed to multiple source of noise that emanates not only from automobiles, factories, aircrafts, trains but frequent social and religious festivals, loudspeakers, radios and so forth adds substantially to the existing level of noise pollution. Noise is sometimes annoying and often it interferes with ongoing activities. It may cause discomfort and can be physiologically and psychologically harmful.

Properties of Noise

Some of the main characteristics of noise are intensity, frequency, duration, predictability, and loss of control over the termination of noise. However, most of the behavioural researches have been focused on noise intensity (Evans & Cohen, 1987). The intensity or loudness depends upon the amplitude of the vibration that initiates the noise. The intensity of noise is measured in decibels (dB). The whispering produces a noise of 20-30 dB, normal conversation 60-65 dB, and boiler factories 120 dB and aircraft takeoff 140 dB. People usually have the capacity to tolerate noise up to 85 dB without substantial damage to hearing. The frequency of noise is denoted as Hertz (Hz). A normal human ear has the capacity of hearing frequencies ranging from 20 to 20,000 Hz. However, this range is reduced with age and other subjective factors.

The range of frequencies below 20 Hz is infra audible and those above 20,000 Hz is called ultra-sonic. Many animals have the capacity of hearing sounds that are inaudible to human ear.

Measurement of Sound

A number of instruments are used to measure different properties of noise. Some commonly used instruments in noise related studies are Sound Level Meter, Octave Band Frequency Analyzer, and Audiometer. The Sound level meter measures the intensity of the sound in decibels (dB). The octave band analyzer measures the sound in octave bands. The characteristics of the noise mainly in terms of high/low/variable pitch can be understood by examining the sound spectrum. Finally, the Audiometer measures hearing ability. The zero line at the top of the audiogram represents normal hearing. Noise induced hearing loss shows a characteristic dip in the curve at the 4000 Hz frequency.

Effects of Noise Exposure

Broadly, exposure to noise may result in auditory and non-auditory effects. Two major types of auditory effects are labelled as Auditory Fatigue and Deafness. The auditory fatigue is generally experienced when the noise level is in the region of 90 dB and greatest at 4000 Hz. People usually complain about various side effects such as whistling and buzzing in the ears. A survey conducted by Kumar and Singh (1986) in Kolkata reveals that 8.9 per cent of the population in and around the city are facing hearing problems not because of any physical or hereditary reasons but because of noise pollution. People complain about temporary/permanent deafness when they are continuously exposed to high frequencies and high intensities of noise. People are generally unaware about hearing loss during the initial phases of noise exposure. In a majority of cases temporary hearing loss occur when noise frequencies ranges from 4000-6000 Hz. Usually, exposure to continuous noise of 100 dB or more may cause temporary/permanent damage to ear. The study of Raja and Ganguli (1983) reveal that exposure to 110 dB noise deteriorates the hearing capacity of printing press workers relative to employees working in academic institutes. The study conducted by Singh (1987) reveal that workers employed in noisy environment for more than ten years suffer from acute hearing disabilities. Furthermore, the noise levels between 85-120 dB are harmful. Incidentally, in different industries the noise levels range from 95 to 127 dB (Singh, 1987). Evidence exists that the inner damage may vary from minor changes in the hair cell to complete destruction of the organs of Corti. Exposure to noise above 160 dB may rupture the tympanic membrane and cause permanent hearing loss.

A number of publicized research undertaken by social scientists and medical professionals report serious physiological and psychological damage caused by excess environmental noise in Indian towns and cities. The WHO suggests that noise can affect human health and well-being in a number of ways, including annoyance reaction, sleep disturbance, interference with communication, performance effects, effects on social behaviour and hearing loss. Noise can cause annoyance and frustration as a result of interference, interruption and distraction. Activity disturbance is regarded as an important indicator of the community impact of noise (Australian Environment Council 1988). Furthermore, the EPA (Environmental Protection Agency) of Australia cites research into the effects of noise on human health indicating a variety of health effects. People experiencing high noise levels differ from those with less noise exposure in terms of increased number of headaches, greater susceptibility to minor accidents, increased reliance on sedatives and sleeping pills and increased rates of admission in mental hospital.

Exposure to noise has also been associated with a range of possible physical effects including—colds, changes in blood pressure, other cardiovascular changes, increased general medical practice attendance, problems with the digestive system and general fatigue. The Australian EPA also suggests that there is fairly consistent evidence that prolonged exposure to noise level at or above 80 dB(A) can cause deafness. The amount of deafness depends upon the degree of exposure. Hearing problems such as Tinnitis are also associated with sustained exposure to high level of environmental noise.

Experimental Research on Noise

As early as in 1957, Broadbent reported the adverse effect of intermittent noise on task performance. Later, many studies have reported that a periodic noise degrades performance more than continuous noise (Glass & Singer, 1972). Cohen and Weinstein (1981) reported that although performance is sometimes impaired by noise, it occasionally improves and very often changes. The study of Glass and Singer (1972) demonstrates that a periodic noise does not produce more task deficit than periodic noise. Although, unpredictable noise does not degrade simple task performance, it does exert a greater impact on complex task performance relative to predictable noise. When people are lead to feel that the noise is inescapable and unavoidable and that they are not able to control the outcomes, they become helpless and stop trying. Thus, it is not the stressful event that causes problems to urbanite, rather than the message of helplessness that causes discomfort. This explanation derives support from the studies conducted by others (Sherrod, 1974; Sherrod, Hage, Halpern, & Moore, 1977). Sherrod and his colleagues designed an experimental study to study the effects on noise involving four conditions. In the first condition subjects were given control over starting the noise. The second condition gave subjects control over the termination of noise. The third group was given control over both starting and termination of noise. Finally, the fourth group was not given any control. The result revealed that subjects possessing greater control over the noise did indeed worked for longer duration at insolvable puzzles after the noise had stopped. The result obtained was consistent with the learned helplessness perspective. The study of Glass and Singer (1972) further reveals that the effects of noise are not caused by an inability of people to adapt to it. It is the aftereffect in the form of psychic costs that causes major noise-induced psychological and behavioural problems. Finally, the social and cognitive context in which the noise occurs is important in determining the noise induced effects.

The study of Cohen and Weinstein (1981) reveal that exposure to high intensity (90 dB) and unpredictable noise increases catecholamines, blood pressure, heart rate, and skin conductance. Early studies conducted in the Indian context report deterioration of mental and mechanical task performance at the 80 dB level (Kumar & Mathur, 1969). The study conducted by Bhattacharya, Roy and Chatterjee (1978) on a sample of non-technical subordinate employees report improved performance under the condition of noise but adverse effects on various physiological measures related to pulse rate and blood pressure. A series of studies was conducted to examine the effects of noise variation on mental efficiency and performance (Batra, Muhar, & Bhatia, 1990; Bhatia & Muhar, 1988). Multiple group design was utilized and the subjects were required to multiple tasks under different conditions of noise. Based on the scores obtained on noise sensitivity scale, the pre-selected subjects were required to perform on various multiplications tasks with accuracy and speed under varied noise conditions. Various physiological indices like Galvanic Skin Response (GSR), blood pressure and pulse rate were measured to assess the energy consumed during task performance. The results of this study revealed that under the unperiodic noise condition the performance on mental tasks was adversely affected while the quiet condition was found to be the most favourable for mental task performance. Furthermore, irrespective of noise sensitivity of the group unperiodic noise was found to be the most disruptive. Respondents having high score of noise sensitivity consumed high energy while performing various tasks under noise conditions relative to their counterparts whose scores were low on noise sensitivity. The results of these studies were explained within the perspective of adaptation. Other studies using more or less similar paradigms reported similar adverse effects of noise on task performance and learning (Dwivedi, Mishra, & Tripathi, 1984; Srivastava & Dwivedi, 1990).

Granted the fact that people residing in urban areas are exposed to both crowding and noise, experimental studies were planned to examine the main and combined effects of both these stressors on cognitive tasks performance and effect (Nagar & Pandey, 1987; Nagar, Pandey, & Paulus, 1988). The experimental study utilizing 2 × 3 factor design with two levels of density (high and low) and three levels of noise (High noise, 94 dB Low noise, 50 dB and no noise) was used. Density was manipulated by keeping the room size constant and increasing or decreasing the number of subjects in the room. The noise was tape-recorded and variation of noise was attained by increasing or decreasing the volume of the tape-recorded sound. The results obtained by Nagar and Pandey (1987) revealed that under high density and high noise condition the performance of complex task was deteriorated. However, with the variation of density and noise no effect on simple task performance was found. The significant interaction effect between noise and density on complex task performance was found. The effect of noise variation was constant under the condition of high density, but under the low density condition performance showed continuous deterioration from low noise to high noise condition. The results of this study were explained within the perspective of arousal theory. In the subsequent study crowding experience was incorporated together with the independent variable of density and noise, to investigate their effects on cognitive tasks and affect. The crowding experience scale was constructed and high and low crowding experience subjects were selected with respect to the scores which they obtained on this scale. These pre-selected subjects were exposed to various conditions of density and noise. The results indicated that high experience to crowding subjects when asked to work under the high density condition performed significantly better and exhibited positive affect relative to their low experience to crowding counterparts who worked under high density condition. Prior crowding experience did not influence reactivity to noise as in high noise condition the subjects performance was lower than low and no noise conditions irrespective of their prior crowding experience.

Based on the results one can conclude that prior crowding experience prepares individuals for dealing with high density settings but not necessarily to noise.

Field Studies on Noise

A number of studies were carried out in naturalistic settings like airports. The study of Herridge, (1974) revealed that people living near the Heathrow Airport or Los Angeles International Airport reported higher psychiatric-hospital admission rates relative to their counterparts who were not living near these airports. Consistent results were found earlier by Kryter (1970). Aircraft noise was found to be a potent contributing factor in health disorder requiring psychiatric consultation. In a similar vein, the study of Knipschild (1977) provides evidence that men living near Amsterdam's Schiphol Airport were more likely to be under treatment for cardiovascular disease and high blood pressure relative to others who were not living near the airport. The results of these studies should be interpreted with caution because they failed to control carefully the social class variables. A recent study conducted by Halpern (1995) attempted to examine the relationship between traffic noise and psychosomatic symptoms. He found that with the increase in the levels of noise consistent increase in various symptoms such as nervousness, depression, sleeplessness, undue irritability, and asthma was found.

A number of studies were designed by Cohen and his associates on third and forth grade children attending the noisiest elementary schools in the air corridor of Los Angles International Airport. These children were matched with other children who were studying in quiet schools on major demographic variables like age, social class and race. Both groups of children were intensively studies and tested to study the effects of noise on various dependent variables. The results of these studies revealed that children from noisy schools had higher diastolic and systolic blood pressures relative to children who were studying in quiet schools. Furthermore, children from noisy schools were more likely to fail on cognitive tasks and were more likely to give up the tasks before the allotted time (Cohen, Evans, Krantz, & Stokols, 1980). Similar results were reported in another study conducted on children studying near Orly Airport in Paris. The children who attend schools near airport exhibited less frustration tolerance relative to children who were from quiet school (Cohen & Weinstein, 1981).

In a study involving railway personnel noise was found to be a major contributing factor in increased anxiety, lower self-concept, and low morale. Morin (1989) found that work involvement, family situation and aspects of identity mediated the adverse effects of noise in the supermarkets and cafeterias. In another recent study noise was found to be one of the possible contributing factor linked with miscarriages and birth defects (Veitch & Arkkelin, 1995).

A number of studies on the effects of noise were conduced in a range of urban settings like housing complexes, shopping malls, hospitals (Cohen, Glass, & Singer, 1973; Hopkins, 1994; Halpern, 1995). To study the noise effects Cohen, Glass and Singer (1973) conducted a study on children who were living in the high-rise residential complexes located near the busy New York highway. After controlling for many extraneous factors they found that on average the children who lived on the noisier lower floors had poorer hearing discrimination and lower reading ability as compared to children who resided in the quieter upper floors. Consistent results were reported in other studies where children living near the railway tracks showed poor reading scores relative to children who were far away from the tracks (Bronzaft & McCarthy, 1975). Some studies have examined the relationship between noise and social behaviour like aggression, attraction and altruism. For instance, evidence exists that noise facilitate aggressive behaviour (Green & O'Neal, 1969; Konecni, 1975) and reduce helping behaviour (Page, 1977). Hopkins (1994) conducted a study on noise in the shopping mall and reported that the physical intensity of noise had a detrimental effect on shoppers well-being.

The studies discussed provide sufficient evidence about

the harmful effects of noise. While it is certain that noise cannot be totally eliminated. However, attempts should be made to control the intensity of noise in many places. For instance, to control the noise level several states including New Delhi have adopted legislation and prohibited the use of loud speakers, vehicle horns and other sources of noise. In Industries law is enforced so as to reduce the adverse effects of noise. Workers who are exposed to noise that are louder than 85 dB and above 150 Hz should be protected. Periodic audiogram, checkups and use of ear mufflers are made compulsory for workers who are exposed to noise. To check the noise level in offices and factories sound absorbing materials are used to cover the wall so that the noise level is controlled at source. The government should educate the masses through all available media and solicit their participation to deal with the noise hazard.

Chemical Pollution

Chemical pollution is a by-product of industrial and technological advancement. Most of the chemical pollutants are increasingly finding their way through air and water causing severe ill effects to human, plants and aquatic life. The world has already witnessed the serious consequences of various industrial and technological disasters such as nuclear explosions, leakage of poisonous gases in the atmosphere and oil spills. On repeated intervals large numbers of wildlife and aquatic life have died as a result of spillage of oil. Industrial accidents killing thousands of people are also witnessed all across the world. In all major cities of the world people are suffering from the physiological and psychological ill effects of chemical pollutants which are released in the atmosphere from various sources like factory chimneys and automobiles.

In the 1950s a plastic industry in Japan was banned. Evidence exists that the release of mercury in the bay by the industry was consumed by fish which eventually reached the people resulting in the death of 43 people, severely disabling 111 and affecting 19 babies with congenital diseases and deformities (Veitch & Arkkelin, 1995). Another example of chemical pollution and its severe ill effects on humans occurred in the residential area located in the New York State. A dwelling built on the site of a chemical dump was found to be one of the salient causes responsible for the incidence of miscarriage, blood and liver abnormality and birth defects (Veitch & Arkkelin, 1995). The DDT once considered as a wonder pesticide is today banned in many countries including India because apart from various negative consequences including sex related disorders. Furthermore, this pesticide continues to have polluting effect for up to 100 years after it has been used (Cassidy, 1997).

A number of lessons summarized below can be learned with regard to the ill effects resulting from chemical pollution (Veitch & Arkkelin, 1995). These are:

1. Our society is producing chemicals at a rate, which outstrips our ability to determine their health and environmental impact.
2. We have not identified appropriate means for the disposal of many chemicals.
3. The effects of chemicals are subject to individual differences.
4. Pollutants may have thresholds, which means that no effect is seen until the threshold is crossed, and the effects, may be fast and furious.
5. The chemical form of the pollutant has a major effect on its toxicity, in that the form of the chemical being disposed of may be fairly harmless, but in reaction with other chemicals either naturally in the environment or from some other source, they may become lethal.
6. Some effects of the pollutants are reversible, but some are not.
7. The effect and level of a pollutant may change as it passes through the food chain. In Minamata bay the mercury didn't kill the fish, but it did kill humans.

(From Veitch & Arkkelin, 1995, pp.133-134)

Evidence exists that even if the chemical pollution has no

direct effect on physical health it is responsible for various psychological consequences. The report of David and Wessely (1995) on the aftermath of an accidental dumping of 20 tons of aluminium sulphate into water supply provide no evidence of any long-term health effect. However, many people complained about somatic symptoms and increased levels of anxiety.

A number of researchers have linked chemical pollution with deterioration of cognitive performance in children (Halmiova & Potasova, 1994; Potasova & Arochova, 1994; Potasova, 1992). A Group of children from an industrially polluted environment was compared with matched controls from a non- polluted environment on a range of memory tasks. The results of these studies indicated that children belonging to polluted environment had poorer short-term memory, reduced attention span and poorer sensory motor skills relative to their counterparts from the non-polluted environment. Similarly, exposure to high levels of lead in the environment was linked with the impairment of cognitive functioning in children (Raab, Thompson, Boyd, & Fulton, 1990; Thompson, Raab; Hepburn, & Hunter, 1989). Furthermore, based on extensive literature review Spittle (1994) reported a strong link between exposure to high levels of fluoride and impaired CNS functioning in terms of memory and cognitive deficits.

A recent study conducted in the industrial area of Calcutta by Chattopadhyay, Som, and Biswas (1993) found evidence that exposure to air pollution causes both psychological and physical distress in adults. Another study conducted on adults found evidence of pollution induced deficit in reaction time and concentration (Bullinger, 1989). Apart from health related problems and cognitive impairment researchers have reported that exposure to air pollution is linked with affective disorders in adults. For instance, the study conducted one year after the Exxon Valdez oil spill by Palinkas, Petterson, Russell, & Dows (1993) on members of Alaskan community found evidence of a breakdown in social and family relationship, increased drinking and family violence and health complains.

The psychological effects of chemical pollution are very

often not incorporated in diagnosing psychiatric cases (Cassidy, 1997). Questions regarding exposure to chemical pollution and toxin substances should be incorporate in the clinical interviews to overcome or at least reduce the possibility of misdiagnose by psychiatrists (Dumont, 1989).

In short, chemical pollution has become part of our modern lifestyle. The physical and psychological consequences of chemical pollution on humans, plants, animals and aquatic life is immense. Environmental psychologists may play important role in exploring the various psychological interventions and prevention strategies so that adverse affects are at best mitigated.

Air Pollution

In recent years air pollution has become a major source of stress. Air pollution may be conceptualized in terms of the presence of one or more particulate in the atmosphere for such duration and quantity that is injurious to human health and welfare. It is widely believed that air pollution threatens our psychological well-being, reduces adaptability, inefficiency and enhances the health problems. Researchers have tried to determine whether or not there is a causal relationship between exposure to various types of air pollutants and health effects? Another pertinent question relates to the behavioural and psychological consequences of air pollution. Researchers are interested to examine how people appraise the pollution related threats and how such adverse effects are reduced. The behavioural studies on air pollution on human are less in number as compared to studies conducted on other environmental stressors like crowding, noise and heat. It has been pointed out that very few studies have explicitly examined whether or not the environmental conditions that causes air pollution are stressors (Evans & Cohen, 1987). Behavioural research covering various psychological aspects is at very early stage and only suggestive inferences can be drawn.

Toxic air pollutants are poisonous substances in the air that comes from natural sources (e.g. radon gas coming up from the ground) or from other range of manmade sources like emission from factories, automobiles. These pollutants adversely affect the environment and put us at health related risks. After inhaling toxic pollutants some health problems occur very soon like watery eyes. Other health problems may be serious and life threatening and their effects are seen after a long period of time like lung damage, cancer. For example, people living near the factory that releases cancer causing chemicals inhale toxic air and are at risk of getting cancer. Breathing toxic air could also contribute to emphysema or reproductive disorders (EPA, 1991).

Recently, the U.S. EPA's Toxic Release Inventory 2000 revealed that the 7.1 billion pounds of chemicals released into the environment by American industries, 1.9 billion pounds, or 27 per cent of overall discharges, were released into the air. 58 per cent of all industrial air emissions were reported by the manufacturing industries, while electrical utilities accounted for 41 per cent (EPA, 2000). The number of deaths associated with air pollution in U.S.A. ranges from 50,000 to 100,000 per year (Dockery & Pope 111, 1994). Furthermore, Congressional Testimony given by George Thurston (1997) reveal that for every 75 deaths per year due to air pollution, there are estimated 265 hospital admission for asthma and 240 non-asthma respiratory admission, 3500 respiratory emergency doctor visits, 180,000 asthma attacks, 930,000 restricted activity days and 2,000,000 acute respiratory symptom days. Furthermore, the US based, Environmental Media Services (2003) has narrated report which is based on the results obtained from a 16 years study published in the Journal of American Medical Association. The results of this reveal that in many metropolitan areas residents are at high risks of dying from lung cancer and heart disease because of their long-term exposure to fine particulate air pollutants. Furthermore, exposure to fine particles emitted by coal-fired power plants, factories and diesel trucks and risk associated of lung cancer is comparable to the risk posed by long-term exposure of second-hand smoke from cigarettes. The U.S. scientist reported evidence in February 2002 that ground-level ozone is a "causative factor" in development of childhood asthma. In high-ozone cities, children who played the most outdoor sports were 3 to 4 times more likely to develop asthma relative to kids who played no sports. Thus, the results suggest that ozone pollution is a contributing factor causing higher asthma rates (EPA, 2002).

The air pollution level of India's capital city, i.e. New Delhi is very high as compared to other important cities and therefore, health hazards of citizen as well as visitors of New Delhi are at danger. Delhi is rated as one of the most polluted cities of the world and more and more Capital's residents are subjected to the adverse affect of deadly pollution (Press release, 2001). The automobile, industry and domestic sectors are contributing significantly to the increase in air pollution levels. At all the traffic junctions and market places monitored air was reported to be of poor quality. Air samples even from school premises have presence of sulphur dioxide, nitrogen oxides and suspended particulate matter much above permissible limits. "Cars also throw out fine-grained particulate, which because of their small size are able to penetrate deep into the lungs. The damage airborne soot and exhaust pollution wreak on human lungs and the heart is noted prominently by media." In Delhi there are 1,26,000 industries [small and medium scale units (Government of National Capital Territory of Delhi, 1999)]. Out of which 98,000 are categorized as non-conforming the Master Plan of Delhi (Ministry of Environment and Forests, 1997).

The SPM (suspended particulate matter) in Delhi's air exceeds the permitted levels by over 100 per cent. The RPM (respirable particulate matter) levels were also found to be very high at most of the monitoring sites of Delhi. The prominent sources of RPM are man-made (dust settled on the road side, auto exhaust, coal combustion). The RPM constitutes nearly 40 per cent of the SPM in the ambient air of Delhi which includes gaseous pollutants like SO_2, NO_x, Benzene and Carbon monoxide (Urban Transport and Environment, 2002). Besides outdoor air pollution, indoor air pollution from wood and coal stoves is killing 600,000 women per year is noted by Desai,

United Nations Under-Secretary General for social and economic affairs (*Times of India,* 17 Feb., 2002).

The study of Evans, Jacobs and Frager (1982) provide evidence that exposure to carbon mono-oxide increases heart rate to fulfil the body's demand of more oxygen. The results obtained reveal that carbon monoxide tend to disrupt the performance on multiple attention tasks and also interferes with tasks that requires sustained attention. While insufficient research data on carbon monoxide and memory exist, but there is some evidence that carbon monoxide may slow down memory. Based on the results cited in various reviews one can conclude that air pollution causes respiratory problems in humans (Coffin & Stokinger, 1977; Goldsmith & Friberg, 1977).

Granted the fact that chemical pollution invades our physical and psychological world and reaches us through the air we breathe the water we drink and through food we eat. A number of studies have measured people's attitudes and awareness of environmental pollution (Pathak & Vohra., 1997; Evans Jacobs, & Frager, 1982). When directly inquired about air pollution people report their annoyance. However, when respondents are asked to rank order community problems, air pollution is not considered important relative to other community problems (Barker, 1976). The study of Barker (1976) suggest that awareness and concern are greater for those who are more educated, women, and for people whose scores are high on internal locus of control. Finally, people who have resided for a long times in areas where the quality of the environment is poor or people who are economically weak and their livelihood is dependent on a major pollution exhibit less bothered by air pollution. A recent study conducted by Pathak and Vo..a (1987) on university students attempted to assess the level of awareness, attitude and behavioural commitment regarding twelve major environmental issues. The results of this study reveal that of all the twelve issues, the highest level of awareness was regarding the issues pertaining to pollution by motor vehicles. As compared to males the female subjects reported more concern and exhibited more behavioural commitment towards environmental issues. The results of this survey are only suggestive because the size of the sample was very small and the data was collected on highly educated postgraduate students from Delhi. More studies on diverse population with large sample size are needed to draw any firm conclusion.

There are a few studies linking air pollution with interpersonal relationship and emotional states (Rotton, 1983; Evans , Jacobs; & Frager, 1982). The study conducted by Rotton *et al.* (1983) provide evidence that both malodours and smoke of cigarette in the immediate environment enhances annoyance and adverse effects the emotional states of the individual. Furthermore, exposure to poor air quality diminishes the interpersonal relationship so long as the target person is not viewed as also suffering from noxious air. However, when both the target person and the respondents are exposed to similar poor air quality empathy is enhanced for the other person. Finally, some studies have reported evidence of low empathy under the poor air quality conditions and increased aggressive tendencies in conditions having elements of increased air pollution (Evans & Jacobs, 1982).

It is very well accepted that inhaling smoke from tobacco whether directly or passively is injurious to health. The study of Hepper (1992) reports that passive smoking is more detrimental to health relative to active smoking because a major part of the smoke inhaled is unfiltered, while smokers sucks it through a filter. Evidence exists that children and partners of smokers are more prone to respiratory problems and lung cancer (Veitch & Arkkelin, 1995). The study conducted by Jarvis, Foulds, and Feyerabend (1992) reports that the levels of nicotine in the saliva of bar staff was equivalent to those people who are heavy smokers. Some studies have also demonstrated that reducing the amount of cigarette smoke in the home reduces the risk of asthma and respiratory problems in children (Greenberg, Strecher, Bauman & Boat, 1994). Keeping the public interest in mind the government has imposed ban on smoking cigarettes in many public places. In India, with the growing public awareness and government's action more and more displays requesting people and sometimes prohibiting smoking in many public places including airports, railway station, libraries, offices and factories are easily noticed which is a positive indication.

Adaptation and habituation effects to air pollution are also examined by some researchers. Some studies suggest that exposure to chronic air pollution generate feelings of helplessness and hopelessness (Evans, Jacobs & Frager, 1982). It seems that people do not feel like complaining about air pollution even when they are annoyed largely because of their feeling that it will solve any purpose. In a similar vein, exposure to uncontrollable air pollution is linked with performance aftereffects is also noted by researchers. Finally, the study of Goldsmith and Friberg (1977) provide some evidence of short-term habituation in respiratory sensitivity to air pollution. Furthermore, exposure to chronic air pollution enhances inhibition of immunological response and increases the tendencies of cardiovascular disease.

The overall picture that emerges out seems to suggest that the stress related effects are at least empirically not strong. Although, some evidence exists that air pollution is linked with health problems, performance deficit and negative affective states. Supportive data are also available, showing habituation and adaptation with chronic exposure to air pollution. However, to draw firm conclusion on various psychological and behavioural aspects of air pollution more systematic research investigations are needed.

Summary

Based on the available evidence it seems that environmental pollution have effects on human health and well being that can be characterized as stress effect. However, for air and chemical pollution the stress-related effects are relatively weak as compared to the effects of noise on human behaviour. One reason may be that the quantum of research work done on various aspects of noise is more relative to air and chemical pollution. The importance of psychological factors and their potential to predict various outcomes are noted in the context of models dealing with control, arousal and behavioural constraint (Baum & Paulus, 1987).

A number of gaps exist in the broad area of environmental stress. For instance, a better conceptual understanding of the relationship between stress and cognition is needed. More studies are needed to understand how cognitive processes are affected by ambient stressors. Granted the fact that people are simultaneously exposed to multiple environmental stressors (e.g. noise, crowding, pollution), studies incorporating multiple stressors in a single research tradition are not adequate. Only few studies have incorporated multiple stressors (crowding and noise) to examine the additive and multiple effects on various outcome measures (Nagar, 1998; Baum & Paulus, 1987). More systematic studies are needed to examine the effects of multiple stressors on human behaviour. In addition, more research are needed to develop a better perspective on stress and its ill effects by examining the issues pertaining to how an individual exposed to stressors develops tolerance to handle the adverse effects of stressors, what are the limits of tolerance and how costly is the adaptation.

Another gap in this area is lack of appropriate indices of stress. With the exception of a few crowding studies (Evans & Cohen, 1987), there has been essentially no work on verbal and non-verbal indices of stress during exposure to other environmental stressors like noise and pollution. Research on adaptive processes and environmental stressors has focused primarily on habituation, performance after effects, and indices of health status. However, there has been very little research on how coping with one stressor affect our ability to deal with another source of environmental demands.

With regard to noise and air pollution it is also very pertinent to explore the environmental parameters and situational factors/contexts that are more likely to produce negative effects. We do know that ambient stressors are capable of producing negative affects. We also know that some degree of stimulation is essential to concentrate properly on various mental tasks and physical work. However, in order to enhance our understanding it is essential to know at what level of noise and air pollution the individual can perform optimally.

Apart from the gaps existing in the literature of environmental stress there is also a growing concern among psychologists that their research is not taken seriously by the policy makers. A framework presented by Ruback and Innes (1988) suggest two dimensions along which the relevance and irrelevance of psychological research can be considered: (a) the number of policy variables investigated and (b) the extent to which the dependent measures are useful to the practitioners or policy makers. In order to make the psychological research practically relevant the need is to select those dependent measures that have higher utility and are measured not at the individual level but at the system level? In addition, selection of the policy variables must be clearly operationalized so as to facilitate the policy makers what should be changed. The need is to communicate the implications of research findings to policy makers as clearly and directly as possible. Thus, environmental research should be conducted in a naturalistic environment taking both policy variables that can be changed and dependent variables that are of high practical utility. Pursuing the research of this order will not only enhance the visibility of the discipline of psychology but will create global awareness about the environmental issues affecting the adverse relationship between man and his physical environment.

REFERENCES

Agarwal, A., & Narain, S. (eds.). (1984-85). *The State of India's Environment: A Citizens' Report* (Second), New Delhi: CSE.

Asthana, H., & Jain, U. (1986). Alienation in Bhopal MIC gas victims. *Paper Presented At XI World's Congress of Sociology*, New Delhi: August.

Barker, R. (1976). Planning for environmental indices: Observer appraisal of air quality. In K. Craik & E. Zube (eds.), *Perceiving Environmental Quality* (pp. 175-204). New York: Plenum.

Batra, A., Muhar, I., & Bhatia, P. (1990). Environmental noise and mental work. *Indian Journal of Psychology*, 65, 45-52.

Baum, A., Fleming, R., & Davidson, L. (1983). Natural disaster and technological catastrophe. *Environment and Behaviour*, 15, 333-354.

Baum, A., & Paulus, P. (1987). Crowding. In D. Stokols & I. Altman (Eds.), *Handbook of Environmental Psychology* (Vol. 1, pp. 553-570), New York, John Wiley.

Baum, A., Singer, J.E.,& Baum, C. (1982). Stress and the environment. In G.W. Evans (ed), *Environmental stress* (pp. 15-44), New York: Cambridge University Press.

Bhatia, P., & Muhar, I. (1988). Noise sensitivity and mental efficiency. *Psychologia*, 31, 163-169.

Bhattacharya, S.K., Rao, A., & Chatterjee, S. (1978). Comparison of motor-manual performance in quiet and noisy environments. *Indian Journal of Medical Research*, 68, 306-311.

Broadbent, D. (1957). Effects of noise on behaviour. In C.M. Harris (ed.), *Handbook of Noise Control*, New York: Academic Press.

Bronzaft, A.L., & McCarthy, D. (1975). The effects of elevated train noise on reading Ability. *Environment and Behavior*, 7, 517-527.

Bullinger, M. (1989). Psychological effects of air pollution on healthy residents: A Time Series approach. *Journal of Environmental Psychology*, 9, 103-118.

Campbell, J. (1983). Ambient stressors. *Environment and Behaviour*, 15, 355-380.

Caspi, A., Bolger, N., & Eckenrode, J. (1987). Linking person and context in the daily Stress process. *Journal of Personality and Social Psychology*, 52, 184-195.

Cassidy, T. (1997). *Environmental psychology: Behavior and Experience in Context. Psychology*. Press Publisher, UK.

Chattopadhyay, P.K., Som, B., Biswas, D. (1993). Air pollution and health hazards: An exploratory study. *Indian Journal of Clinical Psychology*, 20, 25-30.

Coffin, D., & Stokinger, H. (1977). Biological effects of air pollutants. In A.C. Stern (ed.), *Air Pollution* (3rd ed. Vol. 3, pp. 231-360), New York: Academic.

Cohen, S. (1978). Environmental load and the allocation of attention. In A. Baum, J.E. Singer, & S. Valins (eds.), *Advances in Environmental Psychology* (Vol.1). Hillsdale, NJ: Lawrence Erlbaum Associates Inc.

Cohen, S., Glass, D.C., & Singer, J.E. (1973). Apartment noise, auditory discrimination and reading ability in children. *Journal of Experimental Social Psychology*, 9, 407-422.

Cohen, S., & Weinstein, N. (1981). Non-auditory effects of noise on behaviour and Health. *Journal of Social Issues*, 37, 36-70.

Cohen, S., Evans, G., Krantz, D.S., Stokols, D.(1980). Physiological, motivational, and cognitive effects of aircraft noise on children: Moving from the laboratory to the field. *American Psychologists*, 35, 231-243.

David, A.S., & Wessely, S.C. (1995). The legend of Camelford: Medical consequences of a water pollution accident. *Journal of Psychosomatic Research*, 39, 1-9.

DeLongis, A., Coyne, J., Deakof, G., Folkman, S., & Lazarus, R.S. (1982). Relationships of daily hassles, uplifts, and major life events to health status. *Health Psychology,* 1, 119-136.

Dockery, D.W. , & Pope 111. (1994). Acute respiratory effects of particulate air pollution. *Annual Review of Public Health,* 15, 107-132.

Dumont, J. (1989).Validity of multidimensional scaling in the context of structured conceptualization. Special issue. Concept mapping for evaluation and planning. *Evaluation and Program Planning,* 12, 81-86.

Dwivedi, C.B., Mishra, R.K., & Tripathi, B. (1984). Noise induced arousal in sentence processing. *Perspectives in Psychological Researches,* 7, 24-25.

EPA (2000). *Toxic Release Inventory 2000,* Public Data release, USA.

EPA (March, 1991). *Risk Assessment for Toxic Air Pollutants: A Citizen's Guide,* 450/3-90-024.

EPA (Feb, 2002). Associated press story on *Air Pollution and Health Studies.*

EMS (April, 2003). *Air Pollution and Health Studies.* Washington Post Study.

Evans, G., & Cohen, S. (1987). Environmental stress. In D. Stokols, & I. Altman (Eds.), *Handbook of Environmental Psychology,* (Vol 1, pp. 571-610), New York: John Wiley.

Evans, G., Jacobs, S.V., & Frager, N. (1982). Behavioral response to air pollution. In A. Baum & J. Singer (Eds.), *Advances in Environmental Psychology* (Vol. 4) pp. 237-270. Hillsdale, NJ: Erlbaum.

Everest, L. (1986). *Behind the Poison Cloud.* Hyderabad. Banner Press.

Frankenhaeuser, M., & Johansson, G. (1986). Stress at work: Psychological and Psychosocial aspects. *International Review of Applied Psychology,* 35, 287-299.

Glass, D.C., & Singer, J.E. (1972). *Urban Stress.* New York: Academic Press.

Goldsmith, J., & Friberg, L. (1977). Effects of air pollution on human health. In A.C. Stern (ed.), *Air Pollution* (3rd ed, vol 3, pp. 458-610). New York: Academic.

Green, R.G., & O'Neal, E.C. (1969). Activation of cue-elicited aggression by general Arousal. *Journal of Personality and Social Psychology,* 11, 289-292.

Greenberg, R.A., Strecher, V., Bauman, K., & Boat, B. (1994). Evaluation of a home based intervention programme to reduce infant passive smoking and lower respiratory illness. *Journal of Behavioural Medicine,* 17, 273-290.

Halmiova, O., & Potasova, A. (1994). The effects of environmental neurotoxins on performance of children in sensorimotor tasks. *Studia Psychological,* 36, 103-111.

Halpern, D.S. (1987). Architectural preferences and mere exposure effects: The role of recognition in attitudinal enhancement by exposure: Cited in D.Halpern (1995), *Mental Health and the Built Environment.* London: Taylor & Francis.

Halpern, D. (1995). *Mental Health and the Built Environment.* London: Taylor & Francis Hazarika, S. (1987). *Bhopal: The Lessons of a Tragedy.* New Delhi: Penguin Books.

Hepper, P. (1982). Smoking, passive smoking and smell. *Medical Science Research,* 20, 265-266..

Herridge, C.F. (1974). Aircraft noise and mental health. *Journal of Psychomotor Research,* 18, 239-243.

Hopkins, J. (1994). Orchestrating an indoor city: Ambient noise inside a megamall. *Environment and Behaviour,* 26, 785-812.

ICMR (1986). *Health Effects of Exposure to Toxic Gas at Bhopal.* New Delhi.

Jain, U., & Palsane, M.N. (2004). ICSSR survey in Psychology, Vol. 3, New Delhi: Sage.

Jarvis, M.J., Foulds, J., & Feyerabend, C. (1992). Exposure to passive smoking among bar staff. *British Journal of Addiction,* 87, 111-113.

Knipschild, P. (1977). Medical effects of aircraft noise. *International Archives of Occupational and Environmental Health,* 40, 185-204.

Konecni, V. (1975). The mediation of aggressive behaviour: Arousal level vs. anger and Cognitive labeling. *Journal of Personality and Social Psychology,* 32, 706-712.

Kryter, K.D. (1970). *The Effects of Noise on Man.* New York: Academic.

Kumar, P., & Mathur, C.N. (1969). Neuroticism as a factor in noise distractibility. *Journal of Indian Academy of Applied Psychology,* 6, 27.

Kumar, P., & Singh, J. (1986). Deteriorating urban environment: A case of Calcutta. In K.K. Dubey & A.K. Singh (eds.), *Urban Environment in India* (pp. 389-401), New Delhi: Inter India Publications.

Lazarus, R.S. (1966). *Psychological Stress and the Coping Process.* New York: McGraw Hill.

Lazarus, R.S., & Cohen, J. (1977). Environmental Stress. In J. Wohlwill & I. Altman (Eds.), *Human Behavior and Environment* (pp. 91-127), New York: Plenum.

Lazarus, R.S., & Folkman, S. (1984). *Stress, Appraisal and Coping.* New York: Springer. Morin, M. (1989). Applied social psychology and noise in work situations. *Cahiers Internationaux de Psychologie Sociale,* March, 1, 31-51.

Mason, J.W. (1968). A historical review of the stress field. *Journal of Human Stress,* 1, 6-12.

Misra, G. (1988) *Psycho-Social Competence in Children in MIC Exposed Areas of Bhopal.* Paper presented at the workshop on developmental norms of children,

National Institute of Public Cooperation and Child Development, New Delhi.

Nagar, D. (1998). *Human Reactions to Crowding.* Jaipur: Printwell Publishers.

Nagar, D. (1999). Environmental Stress. In G. Misra (ed), *Stress and Health.* (pp. 238 -262), New Delhi: Concept.

Nagar, D., & Pandey, J. (1987). Affect and performance of cognitive task as a function of Crowding and noise. *Journal of Applied Social Psychology*, 17, 147-157.

Nagar, D., Pandey, J., & Paulus, P. (1988).The effects of residential crowding experience on reactivity to laboratory crowding and noise. *Journal of Applied Social Psychology*, 18, 1423-1442.

Page, R.A. (1977).Noise and helping behaviour. *Environment and Behaviour*, 9, 559-572.

Palinkas, L.A., Petterson, J.S., Russell, J., & Dows, M.A. (1993). Community patterns of Psychiatric disorders after the Exxon Valdez oil spill. *American Journal of Psychiatry*, 150, 1517-1523.

Pathak, A.M., & Vohra, S.S. (1997). Environmental pollution: Awareness, attitude and behavioral concomitant among youth. In U. Jain (ed.), *Perspectives on Environment and Behaviour*, pp. 138-161, New Delhi: Har-Anand Publications.

Potasova, A. (1992). Serial position effect: Indicator of environmental-toxic injury? *Studia Psychologia*, 34, 127-135.

Potasova, A., & Arochova, O. (1994). Effects of toxic agents in the environment of Mental functions in children: Some methodological approaches and findings. *Ceskoslovensha Psychologie*, 38, 131-142.

Raab, G.M., Thompson, G.O., Boyd, I., & Fulton, M. (1990). Blood lead levels, reaction time, inspection time and ability in Edinburgh children. *British Journal of Developmental Psychology*, 8, 101-118.

Raja, S., & Ganguli, T. (1983). Impact of exposure to noise on hearing acuity of employees in a heavy engineering industry, *Indian Journal of Medical Research*, 78, 100-113.

Rotton, J. (1983). Affective and cognitive consequences of malodorous pollution. *Basic and Applied Social Psychology*, 4, 171-191.

Ruback, R., & Innes, C. (1988). The relevance and irrelevance of psychological research: The example of prison crowding. *American Psychologists*, 683-693.

Sethi, B., Sharma, M., Trivedi, J., & Singh, H. (1987). Psychiatric morbidity in patients attending clinics in gas affected areas in Bhopal. *Indian Journal of Medical Research*, 86, 45.

Seyle, H. (1956). *The Stress of Life*. New York: McGraw Hill.

Sherrod, D. (1974). Crowding, perceived control and behavioral after effects. *Journal of Applied Social Psychology*, 4, 171-186.

Sherrod, D., Hage, J.N., Halpern, P., & Moore, B. (1977). Effects of personal causation And perceived control in response to an aversive environment: The more control the Better. *Journal of Experimental Social Psychology*, 13, 14-27.

Singh, P. (1987). *Ecology of Urban India* (Vol. 2, pp. 12-13). New Delhi: Ashish Publishing.

Spittle, B. (1994). Psychopharmacology of fluoride: A review. *International Clinical Psychopharmacology, 9*, 79-82.

Srivastava, A.K., & Dwivedi, C.B. (1990). Effects of noise and intentional processing retention. *Perspectives in Psychological Researches,* 13, 1-7.

Tandon, S.K. (1991). Psychiatric complications of disaster. *Paper presented at the National Seminar on Environment and Behaviour,* Barkatullah University, Bhopal.

Thompson, G.O., Raab, G.M., Hepburn, W.S., & Hunter, R. (1989). Blood lead levels of children's behaviour: Results from the Edinburgh Lead study. *Journal of Child Psychology and Psychiatry and Allied Disciplines,* 30, 515-528.

Thurston, G. (May, 1997). *Congressional testimony.*

Veitch, R., & Arkkelin, D. (1995). *Environmental psychology: An Interdisciplinary Perspective.* Englewood Cliffs, NJ: Prentice-Hall.

Weir, D. (1987). *The Bhopal Syndrome.* San Francisco, Sierra Club Books.

PART 3

APPLICATIONS OF ENVIRONMENTAL PSYCHOLOGY TO COMMUNITY PROBLEMS

7

THE BUILT ENVIRONMENT

ARCHITECTURAL FACTORS AND SOCIAL BEHAVIOUR IN HOUSING

This chapter highlights some intriguing issues concerning the ecological realities of residential design and its effects on a range of social behaviour. Usually, people are not aware or are not concerned with the influence of built environment and its social, psychological and behavioural implications. The potency of various physical factors of the built environment is consciously noticed only when physical or mental harm or discomfort has occurred. Although, people have always been interested in the physical surroundings, the active research interest on the influence of our designed environment on social and behavioural responses as a pertinent area of empirical investigation was identified in the 70s. In this chapter the focus will be on various types of residential environment like dormitories and the high-rise housing. While examining various types of built environment attempt will also be made to examine how some prominent physical and social features of the designed environment interact with behavioural processes so as to promote or inhibit the user's need and expectations.

Overview of Architectural Influence on Behaviour

In earlier days the behavioural effects of architectural and design were not given due importance and its influence was ignored to a large extent. The design professionals prominently emphasized the aesthetic features and rarely bothered how the features of the built environment influence the behaviour of its users. However, for the last three decades or so more behavioural oriented perspectives are integrated into the design process. An awareness regarding the influence of design on the quality of life is gradually evolving among the design professionals. The three major but different viewpoints regarding the degree and extent of architectural influence on human behaviour are briefly summarized below:

Architectural Determinism

The early conceptualization of architectural determinism believes that the properties of the built environment shape the behaviour of the people living therein. In its extreme form, architecture and design are considered as the only causes of behaviour. This viewpoint presents a very elementary picture and fails to account for the relationship between design and behaviour. This perspective ignores the fact that the transaction between people and the environment is reciprocal. The environment can exert change in people and the user can also influence and change the features of the environment. Furthermore, architectural determinism does not allow for the complex interactions between the physical, social, and psychological factors. On the one hand, sufficient evidence regarding the influence of architectural design on various aspects of social and group behaviour exists. On the other hand, an equally compelling evidence is available that seem to suggest that users need and their ongoing activities modify the influence of design on behaviour (Baum & Paulus, 1987).

Environmental Possibilism

The perspective on environmental possibilism developed by Porteous (1977) posits that environment presents an opportunity as well as it sets limits on behaviour. In response to the environment we examine the choices that determine the degree to which the opportunity is realized or barriers are encountered. As opposed to the deterministic notion (environment determines behaviour), the possibilism approach view the environment as a context in which the behaviour occurs, and the outcomes are jointly determined by the environment and the choices we make.

Environmental Probabilism

Another perspective that falls somewhere in between the determinist and the possibilism positions on design and behaviour is called environmental probabilism by Porteus (1977). The probabilistic perspective assumes that an individual may choose a variety of responses in any environmental situation. However, there are probabilities associated with specific instances of design and behaviour. These probabilities result from both architectural as well as non-architectural factors on behaviour. In other words, in response to environmental context, the probabilities of some behavioural tendencies are more likely, relative to other response options.

The three perspectives discussed above on architectural influences on behaviour provide us an indication that social and psychological phenomena are to some extent influenced by architectural design. Before focusing on studies on residential design and its influence on human behaviour, a brief mention will be made on the design process and the stages in which design ideas are implemented.

The Design Process

A number of interior and exterior elements of design like color of the building, lighting, placement of windows, aesthetic quality of the surrounding heavily influence the users behaviour and mood states is extensively noted by scholars (Baum & Valins, 1977; Stokols & Altman, 1987). The design process involves a general description of how different elements embedded in the architectural design are planned and implemented. Before discussing the various stages in the design process, and how design are planned and implemented, it seems pertinent to briefly mention the concepts of congruence and design alternatives.

Congruence

The construct of congruence refers to the degree of fit between design features and the needs and preferences of the user. If there is adequate match between the form and the function, the design supports the necessary behaviour and the positive outcome is usually seen. Thus, design that facilitates the needs, aspirations and preferences are said to be congruent to the people using them. However, if behavioural options are restricted by the arrangement and design of space, users may feel dissatisfied and will exhibit negative emotional states. In reference to residential setting the concept of congruence is sometimes equated with habitability (Nelson, 1976). Habitability is often conceptualized in terms of how well a particular environmental provision fits the needs of its users. This fit range from basic survival in terms of providing shelter to residents, to designing an environment which promotes safety and security to its inhabitants.

Design Alternatives

Determining the proper design alternative is complex because the process involves consideration of range of factors including social, economic, artistic and cultural. For instance, in a residential setting a number of elements, different scheme of color, provision of lighting or arrangement of open space, all reflect design alternatives. In addition, the design process involves weighing various alternatives on different criterion.

Stages in Design Process

In terms of the social-psychological input to design in the context of person we need to view design as a process comprising of various stages. A number of researchers have outlined and elaborated these stages (Baum, Singer, & Baum, 1978; Cassidy, 1997). For example, according to Baum *et al.* (1987) the five stages involved in the design process are related to the awareness of design alternatives, Selection of behavioural criteria, Testing the feasibility of specific design by employing different research methods and implementing the design process and end product. In a similar vein, Zeisel (1981) has described the design process in five stages: programming, design, construction, use and evaluation. A brief description of these stages is given below:

Programming Stage

In a residential complex attempt is made to foresee the user's needs and to analyse whether the various physical and psychological features embedded in the complex are able to meet the users needs. Furthermore, it is essential to have a knowledge regarding the extent to which the users needs are satisfied in future when the project is completed. Thus, in this stage the needs are identified and the specifics of the need may be elaborated. Information must be obtained on the full range of prospective occupants on relevant dimensions with regard to their needs and expectations. For instance, knowledge pertaining to the details about the number of occupants, the adequacy of the open space, provision of amenities, how the occupants will be utilizing the space, the needs and desire of the occupants, and overall purpose of the space may be needed to make the stay of the residents comfortable and psychologically satisfying. While the potential users may provide relevant information pertaining to their needs and desires. However, it is equally true that users are often unaware about the specific effects of environmental quality, noise, crowding, colour of the building and adequacy of light on their health and well-being. Finally, planners and designers must be open to information and advice from all possible sources.

Design Stage

In this stage information collected at the programming stage is integrated in producing a design that is apt to maximally satisfy particularly the social-emotional aspects of the users need and provide them enough space for privacy regulation.

The purpose of the design may be to enhance the psychological and physical health of the potential users and provide them a secure place free of criminal activities.

Construction Stage

The role of psychological input at the construction stage is minimum after the effective execution of inputs received from the programming and design stages. However, there is always a possibility that some aspects are ignored or overlooked at the first two stages of design process. Furthermore, it is not simple and straightforward to convert an idea into physical realities. In order to have an idea of how the project will look like after completion, computer simulation techniques are frequently used. These computer simulations are useful but it cannot be replicated by the total psychological experience of the environment (Stamps, 1994). Thus, it is essential to continuously monitor the change throughout its implementation and be flexible to respond to important issues that arise. Thus, in response to the problem modification can be carried out in the construction phase.

Use Stage

After the project is completed people occupy the allotted space. You may have experienced the initial excitement on moving into a new house, which may be replaced later by disappointment once you have adapted to the new setting. In many metropolitan cities, depending on the floor you are located in the multistorey building you experience sufficient light/insufficient light, variable air circulation, traffic disturbance, foul air, parking space, water problem and so forth. Research has not adequately targeted this phase, but it is important to identify whether adaptation has occurred before evaluation of the effectiveness of the building is carried out.

Evaluation Stage

A number of studies have focused on post-occupancy

evaluation. According to Cassidy (1997) post-occupancy evaluation is centered generally on three main questions: Does the building effectively serve the purpose for which it was designed?, How does the building affect the behaviour and experience of people who occupy it?, and What lessons can be learned for future designs? The information gained after the evaluation provides feedback to the designers to make necessary changes or modifications in their future projects. Barker and Poe (1980) conducted their study utilizing survey technique to analyze the organizational climate of hospital as perceived by the staff and patients. The respondents were further asked to report the changes in terms of a range of dimensions including changes in behaviour. Finally, assessment of behavioural change was measured by observation technique. The researchers were involved in the hospital project from the design process to its completion and thereafter in the post occupancy evaluation stage. The results of this study revealed that organizational climate had improved for all, with prominent improvement in organizational climate were reported by members of the staff. In terms of behaviour a trend of increase in verbal behaviour was found. Furthermore, the initial reaction of visitors was found to be negative was the only negative feedback reported by researchers. It seems that one of the reason for the success of the project may be attributed to the involvement of researcher at every stage of the design process and feedback received during post occupancy evaluation.

Environmental design is a multidisciplinary process and thus integration of ideas from different perspectives is required to make the end product successful. It is generally observed that the primary concern of the architects is aesthetics while the industrialists emphasize on various aspect of economic viability. Finally, the behavioural scientists and psychologists are more concerned about human behaviour and feelings. In most cases the economic dimensions are higher in the order of priority while the issues pertaining to psychological well-being and emotions are either ignored or are given less importance. A number of studies have noted the importance of humanitarian issues and needs in relation to designed residential environment. For instance, the study of Osmond (1957) provides an excellent example of how creativity in redesigning the space from sociofugal to sociopetal can make an inflexible building more effective. The sociofugal arrangement of space inhibits or restricts social interaction while sociopetal design of space tends to promote or enhance the possibility of social interaction. Sommer and Ross (1958) conducted a study in a nursing home day room. The results of the study reveal that a change from sociofugal to sociopetal arrangement of space improved the quality of life of both patients and staff. However based on the results one should not conclude that sociopetal arrangement of space is always good and sociofugal spatial design is always bad in all situations. In fact, in the hospital setup various aspects of the patients need for privacy should be taken into consideration before making any generalization. For instance, Cherulnik (1993) in his study found that interaction among the patients in a psychiatric hospital enhanced as a function of the combined effects of sociopetal seating arrangements and the provision of partition in the sleeping rooms. Furthermore, need for intimacy was increased by the seating arrangement. It seems that feeling of control is an important variable. If the feature of the design helps people to exercise control over the social interaction then they tend to derive more satisfaction. Thus, the balance between the need for intimacy and the need for isolation is important for us and the environment which facilitates us to exercise autonomy are satisfying for us. Based on the review of relevant literature Winkel and Holahan (1986) have identified factors like spatial layout of the environment, privacy considerations, personal control and independence, sharing information, social relationships within the hospital, and environmental stimulation, which contribute to the prevention of psychological and social problems of psychiatric patients. Before we attempt to examine the role of design features and its impact on behaviour, a brief mention on individual differences and design will be made in the next section of this chapter.

Individual Differences and Design

People react differently to the same levels of environmental stimulus including crowding, noise and pollution. However, it has been noted by many scholars that there are not only differences but also there are similarities between groups of individuals (Kluckhohn and Murray, 1953). Barker and Wright (1955) have pointed out that similarities are largely situation dependent and may prove to be useful if combined with some degree of predictability in the design process. McKechnie (1974) have analyzed individual difference within the framework of personality. He developed an Environmental Personality Inventory (ERI) to identify individual differences in preference for particular type of environment. The ERI comprises of eight distinct factors that relates to measure one's preference for urban versus rural settings and one's need for privacy. While using these measures one should be very careful in deriving conclusion because it is essential to consider the issues pertaining to reliability, validity and situation dependence. Moreover, the dimensions related to environmental personality must be viewed within the context of cognitive styles that are learned and modifiable (Pervin, 1953). Thus, if we consider that stability of behaviour is a function of person, environment and the interaction between person and environment, our predictions about different components of behaviour in the process of design are likely to be more accurate. Craik (1976) have reported three major functions in which application of measures pertaining to individual differences is useful and effective. These functions are related to description, prediction and comparison. In other words measures of individual differences help us to describe people in relation to different environment, they are useful in making comparison between people across and within environments. Finally, these measures allow us to make prediction of future behaviour in particular environments. The ERI measures have shown validity in predicting the possibilities of emigration behaviour of students (Gifford, 1987). Based of the research on individual differences generalizations to specific subgroups are possible in the context of multiple-level analysis approach. However, generalization of results for the entire population is not possible from the measures of individual difference.

Effects of Residential Design

For the purpose of this chapter a built environment of a residential setting can be conceptualized as one that has been designed and formed to a large degree by human active interaction. The most important dimension of the built environment is the provision of space. The influence of residential design and built structures on behaviour can be understood in terms of spatial layout of the buildings, the arrangement and positioning of individual dwelling units, the location of day-to-day facilities and amenities like laundry rooms, mail boxes, recreational areas, stairways and so forth. In addition, the interior and exterior arrangement of space in various areas of residential environment provides the users with different behavioural options. The design features may promote or inhibit certain form of social behaviour. While disregarding the deterministic model, which emphasizes the straightforward cause and effect relationship between elements of residential design and behaviour, most of the contemporary theories of environment and behaviour accept the reciprocal relationship between physical environment and behaviour.

The choice of single-family dwelling, located away from the central city is a preferred choice in North America (Sundstrom *et al.,* 1996). Some of the common properties associated with residential satisfaction are identified as: the importance of adequate space, convenient location for services, and sense of security associated with distance from inner cities, attachment to neighbourhood and people, and environmental support for changing work and family roles. Very few studies have compared different types of housing (e.g. apartment vs. home). The study conducted by Paulus *et al.* (1991; 1996) compared similar population residing in apartments and mobile homes. Because mobile homes are stigmatized as having low and negative image (Wallis, 1981), one might expect negative reactions from the residents. In contrast, because of high density and associated noisiness negative ratings on apartments are expected. However, studies conducted on military families by Paulus *et al.* (1996) revealed no major differences between two types of housing in environmental ratings, harmony and well-being. Apartment residents did rate their housing more favourably in terms of services and facilities provided, compared to mobile homes residents. It seems that residents may have simply chosen the different type of housing because of their differential needs (e.g. privacy, facilities, location) and thus may have reflected the same levels of residential satisfaction. Evidence exists that residents base their housing selection on their needs and values (Rapoport, 1985; Paulus & Nagar, 1987) and hence may derive similar levels of satisfaction with their housing of discrepant quality. In contrast, some studies have indicated that objective factors do influence housing satisfaction (Christensen *et al.*, 1982; Galster, 1987). To the extent that low quality of housing can be a source of stress, housing quality and satisfaction should be related to degree of mental and physical well-being (Caspi, Bolger, & Eckenrode, 1987; Christensen *et al.*, 1982).

The literature provides ample evidence that people are not passive to their surroundings. They can actively react, adapt and even change their environments. On the other hand, the built characteristics of the environment do affect the people in significant ways like fostering or inhibiting the social interaction patterns (Paulus & Nagar, 1987). In other words people and various facets of the built environment interact with ongoing social and psychological processes to determine behaviour. While creating the various types of housing structures the architects and planners place considerable importance to the aesthetic but often the salient needs of the users are either not adequately recognized or are missed. The incongruence between environmental features and users needs and expectations creates residential dissatisfaction and disrupts the normal social behaviours between residents. In addition, there are important social, economic and cultural preferences between housing policy maker programme and those who are the recipient of such policy programmes (Cooper, 1975). The Ittelson, Proshansky and Rivlin (1974) writings reflect that the features of the built environment can be related to the users living therein in each of the following five ways:

Physical Factors

The arrangement of the actual residential physical structures like apartment units, facilities, compartmentalization of space in relation to people utilizing the space. One important factor that is extensively researched relates to examining the relationship between actual physical distance (e.g. physical proximity) between residential units on residential satisfaction and social behaviour.

Functional Factors

The residential design and various structures within the residential environment are constructed so that the users needs are met effectively. The concept of functional distance is conceptualized in terms of the extent to which social contact between residents are likely to develop because of environmental provision. Studies conducted in the high-rise residential complexes reveal that functional proximity between apartments aid in social contact and is a deterrent of criminal activities (Newman, 1972).

Cognitive Factors

The residential environment conveys a set of symbolic meaning about the status, values and lifestyles of its residents. The specific dwellings and entire residential environment have great symbolic, affective and instrumental significance to the residents (Rapoport, 1985). The built environment is conceived by Rapoport (1985) as a form of non-verbal communication in which the architects is the 'encoder' of meaning and the 'user' is the 'decoder'. Serious psychological and environmental problems may result if the users and the designers do not share the common language or if the intended meaning is not conveyed properly or misinterpreted.

Affective factors

Various features of the built environment have the capacity to arouse as well as generate emotional feelings in its users. Inbuilt features of the housing can promote the feelings of security, safely, crime and residential satisfaction (Newman, 1972). The interactive relationship reminds us that people are active elements of the environment. The role of psychological factors like choice and self-selection of the particular dwelling is helpful in the understanding of the positive and negative reactions in response to the physical environment (Rapoport, 1985; Paulus *et al.*, 1991).

Social Factors

The literature on community perspective to environmental psychology has extensively focused on the adequacy of the designed environment and fulfilment of social needs of residents. The role of various environmental factors that facilitates or inhibits the social contact and group behaviour between residents is amply documented in the literature (Stokols & Altman, 1987; Paulus & Nagar, 1987). The role of social factors in reducing the ill-effects of crowding, noise and privacy regulation is noted by scholars in diverse residential settings (Baum & Paulus, 1987).

Studies on Spatial Factors and Social Behaviour

The first systematic treatment to understand the role of environmental features in determining behaviour can be attributed to Lewin (1935; 1946). However, the research done by Lewin was primarily focused around social and the physical properties of the environment. It was the classic work of Festinger, Schachter and Beck (1950) which stimulated wealth of empirical research on various aspects of the built environment and its relationship with a range of social and behavioural measures. The influence of physical environment and architectural design on interpersonal behaviour, friendship formation and group development was examined on MIT married couples who were residing in two very different types of environmental settings. Strong support for the importance of both physical and functional distance was found to influence the probability of group development and friendship formation among the residents. Sufficient empirical evidence exist that social interaction among inmates was more prominently centered around those houses that were close in terms of physical and functional proximity. For instance, location or orientations of front doors, the positioning of stairs are both design features that can serve to put residents living in the same building at close functional distance. Similar, result was obtained by Priest and Sawyer (1967) in a residential setting at the Chicago University. The distance between rooms was closely related to the students liking for each other.

In another study, Whyte (1956) examined the role of physical factors in determining patterns of friendship. The apartments were grouped in courts. Across the period of time each court developed a particular culture that persisted irrespective of change in the residents. The prominent role of close physical proximity was found to be the potent factor in the enhancement of social contact. The location of the garden, play areas for children promoted social interactions among residents. Furthermore, evidence that proximity influences people's choice of friends was reported in a survey study of four inter-racial housing projects (Deutsch & Collins, 1957). Two housing projects were totally integrated while in the other two housing complex, blacks and whites were separately housed. Their results reveal that proximity or close physical distance was found to be a potent factor in friendship choice independent of race. Consistent results were obtained in other studies, which supported the relationship between proximity and friendship formation (Caplow & Forman, 1950; Merton, 1948). Interestingly, some scholars have reported the positive relationship between unpleasant social relationship and close physical proximity (Ebbesen, Ejos & Konecni, 1976; Kuper, 1953).

Sense of Community

The above discussion suggests that certain features of the residential environment can have a strong impact on the probability of interpersonal interaction among inhabitants. Granted the fact that dwellings exist is neighbourhood and community and that the interdependent relationship between the two seems imminent. Evidence exist that satisfaction with specific aspects of one's dwelling is a strong predictor of satisfaction with one's neighbourhood (Windley & Scheidt, 1982) and that in turn satisfaction with one's home is dependent on being satisfied with one's neighbourhood. It seems likely that individual families will be influenced by the design of their home. Furthermore, the literature on environment and behaviour also suggest that the larger neighbourhood or local community also influences the residents. Halpern (1995) analysis suggests that it is more important to live what one perceives as a "good" community or neighbourhood than in a good home. This argument focuses on the aspects of the physical design that tend to influence the social interaction patterns and the quality of social support one derives from these settings. It is widely accepted that social support, neighbourhood development and the benefits of positively judged interpersonal contact among inmates are helpful in reducing the adverse affects of stressors (Cohen & Syme, 1985; Unger & Wandersman, 1985). Evidence suggests that various physical factors are helpful in the formation of neighbourhood community network, may serve as a deterrent of crime in the neighbourhood (Newman, 1972). While considerable empirical evidence suggests that various features of the environment induces the likelihood of increased social contact that generates favorable reactions from residents.

It appears that prolonged interaction in the neighbourhood, with its associated exchange of resources (e.g. information, favours, timely assistance), may lead to strong sense of attachment to neighbourhood or sense of community (Sarason, 1974). People living in homogeneous neighbourhood for a long period of time are likely to become interdependent and show high level of attachment to places and people in the neighbourhood locality. A strong attachment to place may develop through a cost and benefit analysis done by the residents in which evaluation in terms of degree of social support is made between the present neighbourhood and the alternate neighbourhood (Proshansky, 1978; Stokols & Shumaker, 1981). Thus, the physical environment must create opportunities for social interaction, while maximizing opportunities for privacy. In other words the physical features of the residential environment must enhance an individual's autonomy in regulating the level of social interaction in which they engage.

Research has documented a variety of positive benefits that are associated with a sense of community. People have a desire to affiliate and interact with groups and the outcome of increased contact is mostly positive (Schaver & Buhrmester, 1983; Schachter, 1959). Residents in the neighbourhood share common concerns and personal problems. They may collectively act in local neighbourhood organizations to solve personal as well as neighbourhood problems (Perlman, 1976; Wandersman, Florin, Chavis, Rich, & Prestby, 1985). Community development and supportive neighbourhood may also be helpful in reducing the fear of crime and reducing the actual crime is noted in the literature (Gubrium, 1974; Sundeen & Mathieu, 1976).

Homogeneity

Another factor that appears to play a major role in the development of relations in environments is related to the degree of homogeneity among group members (e.g. degree of similarity with regard to race, professional and socio-economic status, interest, values). Evidence exist that homogeneous neighbourhoods are more likely to develop into close knit communities, provide more effective social support and results in reduced levels of mental illness (Cassidy, 1997). Festinger *et al.* (1950) and Gans (1967) have pointed out the important role of homogeneity in friendship formation in their studies. Other researches have noted the role of homogeneity in housing satisfaction (Hourihan, 1984; Michelson, 1970). It seems that presumption of similarity may increase interpersonal attraction and the desire to interact (Byrne, 1971). Similarity is also important in the maintenance of relationships (Murstein, 1976). It may be related to reduction of social conflict and enhancement of mutually rewarding interactions. Thus in environments where residents are fairly homogeneous, one should observe enhance impact of spatial factors both in the initiation and maintenance of social interaction and group development (Paulus & Nagar, 1987).

In this section attempt will be made to review studies that examines the impact of various features of the housing environment on social behaviour, emotional states and residential satisfaction. The role of residential design in various environmental contexts like dormitories and high-rise will be discussed.

Studies on College Dormitories

The dormitories of the college provide a naturalistic and useful site for examining the role of various types of residential design on certain social, behavioural and affective measures. A series of studies conducted by Baum and Valins (1977) examines a very homogeneous and comparable population residing in architecturally different type of dormitories (corridor *vs.* suit style dorms) with equivalent physical densities. These dorms differed largely on interior architectural arrangement of residents around shared space. The corridor style dorms housed about 34 residents on a floor with 17 double occupancy bedrooms, with a common bathroom and lounge areas arranged along a central double loaded corridor. The interior spatial arrangement of suite style dormitories dispersed people in small four to six person clusters, each containing its own lounge and bath areas. The physical design features of the suit style dorms contain clear-cut physical provisions of private, semi-private and public spaces. The need to use the hallway by a large group of suit residents was minimum because shared bathroom and lounge spaces were inside the suites. Both the suite and the corridor style dorms housed 22 to 34 residents per floor with comparable spatial densities. However, the architecturally induced variations in functional social densities (e.g. number of residents sharing the common areas) had impact on the resident social behaviour. The results obtained from self-report data reveal that as opposed to suit style dorm residents the corridor residents more frequently complained of unwanted social stimulation and interactions with number of residents and strangers in the shared areas like hallways, lounges and bathrooms. The problems of regulating interactions with others appeared greater in corridor design facilities leading to higher level of crowding stress as compared to the residents who were living in suit-style dorms. Since, the suit residents had to share common facilities with a very small number of residents, the effective regulation of social encounters was possible and families with other residents led to group formation within the suit residents.

The resident responses to the questionnaire data provided information regarding the dynamics of differential control over what happened on their floor of corridor *versus* suit-style dorms. The corridor residents reported excessive unwanted contact and interactions with unfamiliar residents and expressed greater avoidance motivation as compared to their suit dorm residents. While the description of dormitory life based on self-report and questionnaire data clearly indicated the influential role of design variables that varied the levels of social densities in determining the resident reactions to other people. Besides, relying on self-report and questionnaire data Baum and Valins (1979) investigated the behavioural differences of dorm residents. They used behavioural mapping techniques to identify the location and to relate behaviours which occurred in suit and corridor design dorm environments. In case of corridor dorms, most social behaviour occurred in hallways. In contrast, the use of lounge was more frequent by suit residents for social interaction. As far as non-verbal behaviour is concerned the corridor residents reflected the tendency of social withdrawal from hallways and preferred their bedroom to interact with friends and neighbours. The suite residents on the other hand preferred to use the lounge for social interaction. The plausible explanation regarding the corridor resident's avoidance of these spaces where interaction was most likely presumably was their attempt to reduce the frequency of social contact experienced. The other reasons behind corridor residents social withdrawal also seems equally plausible. It seems likely that corridor residents perceived their hallway space as inappropriate for non-social activities and thus avoided the hallway space and preferred their bedrooms for pursuing these activities.

Frequent and unwanted interaction with others in the hallway of corridor design dorms led to the tendency of social withdrawal from hallway to bedrooms largely because the hallways failed to provide semi-private space to its residents. The lounges were converted to semi-private spaces and were used by suite residents for they provided the residents amply opportunity to control the shared space. Baum and Valins (1977) conducted a series of laboratory studies to examine the social withdrawal tendency of corridor residents beyond the dormitory environment. The result of this study clearly demonstrated that relative to suit residents the corridor residents reported more stress and exhibited tendencies towards social withdrawal by maintaining greater distances from the confederates, discouraging eye contact and interaction. The presence of confederate in the waiting hall, before the start of the experiment, increased the discomfort level of the corridor residents relative to their counterparts residing in suite style dorms. The results suggest that social withdrawal and stress effects reported by corridor residents persisted across other situations. The Baum and Valins (1977) study has also reported that increased social density in the corridor dorms interfered with resident ability to regulate control over social interaction. The repeated failure to regulate control over environment and social encounters leads to a sense of helplessness. The results of other studies are found to be consistent (Baum, Aiello, & Calesnick, 1978).

The studies conducted by Baum and his colleagues clearly illustrate the fact regarding the importance of spatial arrangement in residential environment. The physical design features of the corridor design dorms reduces the residents ability to regulate control over their surrounding which seems to create various social interaction problems among inhabitants. It also generates stress, social withdrawal and helplessness. These adverse behavioural and psychological effects seem to persist across different situations beyond residential units.

Studies in High-Rise Dwellings

With the increase in the urban population finding adequate housing has become a matter of great concern. To accommodate a large number of people, the best alternative seems to be the high-rise and multistoried complexes. By constructing the high-rise complexes and skyscrapers man has actively created a third viable built environment comprised of various structures that intercedes between users psychological and social life style. Depending on the type and design of the high-rise buildings certain degree of constraints is imposed on the users in various areas of the buildings. Some physical provisions of the high-rise building are related to sharing of common apartment wall, ceiling, floor, and close proximity with neighbours, sharing of common facilities like laundries, stairways, pools, gardens, parking areas, recreational areas. It seems that the physical features of the residential design are apt to influence the social, psychological behaviour of its residents.

The architects and urban planners have traditionally been using various economic, physical and social criteria to construct adequate multi-family housing. However, in some cases the residents rejected the high-rise housing and eventually some of the buildings got dismantled. The dramatic failure of Pruitt-Igoe in St. Louis and Rosengaard in Malmo, Sweden are some examples of outright rejection of multi-storied housing by its residents. It seems, that designers and planners of these buildings were not able to adequately foresee the users needs and aspirations. It seems that if the design is to work effectively the involvement of users who will live in the dwelling is important (Cohen & Syme, 1985; Lawrence, 1984).

Three important issues are important while examining the impact of built environment on human behaviour (Baum and Valins 1977). These are: (1) To examine the congruence between the arrangement of space imposed by the building and the satisfaction of residents needs; (2) To examine the impact of amenities and resources on residents satisfaction and (3) To examine the role of architectural features in promoting or inhibiting social relationship between residents. An attempt will be made to review some of the studies conducted in high-rise and multifamily dwellings.

During 1970s the interest of researchers (Sommer, 1975; Bechtel, 1977) shifted towards the application of social science knowledge to pursue physical environment studies and help professionals to improve upon the design features consistent to the users needs. To enhance the quality of life for low income residents throughout the United States public housing projects were constructed and were replaced by slum colonies. Hartman (1966) conducted a detailed survey of a slum located in Boston known as West-end. The West-end was dismantled and new high-rise buildings were erected and most of the slum residents were guaranteed accommodation in high-rise complexes. The survey reported that most of the West-enders refused to relocate in the new high-rise. Instead, they preferred to house themselves in those areas that were culturally and ecologically similar to slums from where they were uprooted. In another study conducted by Fried (1963) on West-end, the residents reported signs of grieving, depression and even reported evidence of physical symptoms. The planners were successful in replacing the unhygienic and deteriorating conditions of the slum colonies. However, they overlooked the properties of the social slum, the community life where personal identity of the residents, which were clearly tied to the physical location.

The Pruitt-Igoe project in St. Louis Missouri was intensively studied by Rainwater (1966) and Yankey (1976). The presence and absence of various physical features of the residential design caused serious problems to the residents were noted by researchers. The Pruitt-Igoe project housed 12,000 residents in 43 buildings. All the buildings were 11 storied in height. The Architectural digest (1951) praised the design of the building for the optimally utilizing the space to the extent of having practically 'no wasted space'. However, the way in which the residential space was arranged in the hallways and corridors within the buildings of the housing complex discouraged social interaction among the residents and promoted the sense of anonymity. The stairways were so private and sealed off that they facilitated the criminal and antisocial activities. The survey reported that the stairways were used successfully in pursuing false practices ranging from defecation to sexual intercourse besides making the residents an easy crime victim. The requirement of semi-private spaces for achieving social goals was also not adequately met. Rather than encouraging social interaction with neighbours, the design of Pruitt-Igoe had an "atomizing effect' on social networks.

Another prominent source of discontent reported by the residents in the high-rise was related to lack of parental control over the children activities. The design features of the high-rise inhibited the residents to visually observe their children once they were out of their individual apartment units. Thus, the residents were unable to derive social support from their neighbourhood for the complex failed to provide semi-private spaces around which interpersonal relationship might develop (Yankey, 1976). Furthermore, a close examination of data collected on residents of Pruit-Igoe and nearby slum residents was made to assess the degree of satisfaction and dissatisfaction with individual dwelling units and neighbourhood. The results of this study reveal that Pruitt-Igoe residents reported more satisfaction (75%) with their individual dwelling units relative to slum residents (55%). However, 75 per cent of slum residents reported greater satisfaction with their neighbourhood as compared to 53 per cent of the Pruitt-Igoe residents. Looking at the results it seems plausible that planners of Pruitt-Igoe failed to design the complex where the important social demands and needs of the residents were satisfied. In fact the planners might have ignored the social psychological needs of its residents and have intended to accommodate more and more people in the complex and allowing no wastage of space. Newman (1972) in their study noted that crime, vandalism increased in these complexes apart from breakdown of social mechanism. All these factors together contributed to the general decline of the housing complex and its eventual demolition.

The problems of the high-rise complexes is very well documented in the book entitled *"Defensive space: Crime prevention through environmental design"* (Newman, 1972). One of the most serious behavioural consequences of the physical features of the high-rise is related to crime and safety of the inhabitants. The defensive space theory of Newman attempts to explain how creating the physical provisions, facilitates the social interaction and reduces the frequency of crime rate in the high-rise settings. Newman (1972) defines the concept of defensive space as ".... The range of mechanism—real and symbolic barriers, strongly defined areas of influence, and improved opportunity for surveillance—that combine to bring an environment under the control of residents. A defensive space is a range of living residential environment that can enhance the inhabitants quality of life while providing social security for their families, neighbours and friends." The defensible space theory proposed by Newman offers two important guidelines that may discourage criminal activities. First, defensible space facilitates social interaction that hopefully promotes the feeling of interpersonal trust, social support and group cohesiveness. Second, it evolves ways to increase visual surveillance of immediate environment particularly in non-private areas of surroundings. Newman notes that visually deprived areas are more susceptible to crime. Thus, physical design should provide opportunities for residents to visually survey these areas through location of windows, outdoor balconies and so forth.

Newman (1972) compared two New York housing projects located across the street to test his ideas on defensive space. The Van Dyke project exemplifies a traditional high-rise. In contrast, Brownsvielle housing accomplishes many components of defensible space. The population densities of both the housing projects were more or less the same. The results of this survey indicate that the defensible space design of Bronsvielle had two times less robberies relative to Van Dyke housing. In addition to this 72 per cent less repair and maintenance was utilized in Brownsvielle. The results also reveal that of all the physical design features, building height, ranked among the three most important predictor of robbery rate, explaining 10 per cent of the variance. To test the defensible space ideas Newman, Kohn, Franck and Fox (1975) conducted a survey study on four housing projects. They introduced actual physical modifications consistent with the defensible space theory in two projects. To enhance visual surveillance, lightening facilities in the key area was improved. Fences, yards were constructed to clearly demarcate the territories and play areas for children. The results clearly indicated that as compared to the two unmodified housing projects, the modified projects with provision of space, enhanced the safety of the neighbourhood and reduced the victimization rates.

As an extension of the earlier work on defensible space Newman and Franck (1980) conducted a study in moderate-income housing projects in Newwark, St. Louis and San Francisco to examine factors that influenced crime and instability. Their theoretical causal model incorporated several characteristics of housing development, which were treated as independent variables. These were related to physical design variables (e.g. building size, visual accessibility), social characteristics of residents (e.g. income, adult-teen ratio, ownership status) and available security services (e.g. police and guards). The intervening variables were related to attitudes and actions on the part of residents and management. Finally, crime ratio, fear of crime, burglary rate and housing instability were treated as dependent variables. The results of this study indicated that two physical design variables (building size and accessibility) and two social characteristics (income and adult-teen ratio) were found to be the major determinants of all the four community problems. Of all the predictors burglary was predicted by the degree of visual accessibility. Personal crime is most determined by lower income, and teen-adult ratio in the housing complex. Finally, community instability is determined by building size, visual accessibility and low - income.

An interesting pattern emerges when a close inspection over a set of intervening variables that mediate the effect of various features of the housing development on each of the four community problems was made. It seems that if the domain of concern for residents is extended beyond the individual apartment unit to various immediate areas of the neighbourhood, the lesser will be the probability of crime and fear of crime. The researchers have noted that with the increase in building size, there is a decrease in the use and control of space which leads to fear of crime as well as housing and community instability.

The defensible space theory propounded by Newman is subjected to a great deal of criticism. For instance, Mawby (1977) questions the survey conducted by Newman (1972) on various methodological grounds and states that the Brownsvielle and Van Dyke housing projects cannot be compared on all relevant social and demographical characteristics. Other scholars have objected the Newman's usage of territoriality concept as a moderator of defensible space (Taylor, Gottfriedson, and Brower, 1981). In a similar vein, Krupat (1987) have questioned the model of defensible space as deterministic. Despite intense criticism, the research and theory has stimulated a wealth of research on the interaction of architectural features of residential design, social factors and behavioural measures.

It is believed by many scholars that incorporating the ideas of defensible space in the construction of high-rise housing is necessary but certainly not a sufficient condition for reducing the criminal activities in the residential complexes.

(Merry, 1981). Even the best architectural arrangement of space, well-positioned windows and balconies for surveillance are helpful in only some respects to deter the criminal activities. In addition to this provision of semi-private spaces in various areas of neighbourhood is also helpful in building strong ties and trust among the inhabitants. The social fabric and supportive interaction was found to prove helpful in enhancing the residents control over the surroundings. The design of the buildings should be constructed keeping in mind the users needs which seems very pertinent from the research studies. Thus, the merits and utility of defensible space provisions in housing is generally accepted. Moreover, the demographical characteristics of the inhabitants, users needs and priorities with respect to semi-private spaces and physical arrangements of space may also be helpful in enhancing residential satisfaction and reducing the criminal activities.

The Case Study of Chandigarh City

The discussion in the preceding section is centred on various housing projects located in the United States. Various reasons are discussed to explain why some of the housing projects failed dramatically? What aspects of the residential design are more vulnerable to crimes and antisocial tendencies? How the physical aspects of the design be made more user friendly? In this section the purpose is to highlight some of the good and bad features of the Chandigarh city that came in existence in 1950s and was designed from scratch by one of the noted architect, Le Corbuiser.

Brolin (1972) has intensively surveyed the city of Chandigarh. In his survey he has demonstrated many inherent and external design problems, which were largely related to misfit between designers intension and the users need. The planners and policy makers intention to construct a model of Western city seems to have been defeated for the users needs, aspirations and deep-rooted cultural and social values were largely ignored in constructing various features of the designed environment. Some of the major design problems are highlighted so as to provide the readers an idea how the lack of congruence between users needs and designer's intension hurts the feelings of the residents.

The entire planning of the Chandigarh city was based on a set of Western assumptions and Western lifestyle. The misfit is evident between the messages conveyed by the built environment and the ways the people read them. In addition, to this high discrepancy prevails between the arrangements of space provided to satisfy the users needs. Broden while taking to the residents illustrates —"In the old city the people are closer, in Chandigarh, we don't share one another's joys and sorrow." Thus, in spite of the best amenities provided by the built environment (e.g. hot and cold running water, closed sewers, flush toilets, parks, garbage cans) a sense of anonymity prevails among the residents.

In general, the layout of the city was rectangular in nature and was divided into self-contained twenty sectors. To foster or induce the sense of community, the western style of model was adopted and housing was arranged so that people with similar incomes would live in close proximity to one another. The layout plan to segregate people solely on the basis of income and socio-economic status was not appreciated by the people of Chandigarh for it failed to enhance the sense of community based on proximity and similarity. The Indian cultural system is based on kinship ties and family lines and not on economic status.

In western cities a typical sector is so self-contained that a full range of everyday requirements could be met at approaching distances (e.g. school, grocery shops, health centres, repair shops, small supermarkets). Furthermore, the Indian need system was not considered important by the planners and soon the initial planning of shopping areas where initially all types of shops were located were replaced by specialization in each areas. For instance, hardware and plumbing shops in one sector, auto repair and allied shops in another sector and so forth. The reason for this change was to provide the customers the opportunities to bargain from one shop to another to get the best price. Thus it is necessary to have a number of sellers of the same product in close proximity so that the process of bargaining is possible. In the western countries the prices of each commodity is stamped on each item and this simply what one is expected to pay while in India most customers are not ready to pay the initial price at least in traditional markets without bargain.

To keep the environment clean provision of garbage collection were provided at approachable distances to the residents. Again the planners were not able to realize that in Indian culture the edible garbage is not always thrown in the garbage cans. The edibles are thrown outside the houses for wandering cows to eat. Similarly, the inmates according to the intended plan did not use the interior arrangements and facilities of rooms in the individual dwellings. In West, each room differs in terms of its use. For instance, in India the space usage pattern is very different from the western countries. People may use the drawing room to sleep in the night while dining room is multipurpose and is not used exclusively for eating. Brolin (1972) also reported that in many houses curtains are hanged between drawing and dining rooms to provide desired quota of privacy particularly to female family members when the guests are present. The above-mentioned points are few illustrations, but they certainly make a set of points regarding the discrepancy between designer's intention and the users need patterns. Understanding the demographical characteristics, cultural values and users need is very essential for the designers in advance to achieve the best fit between environmental provisions and residential satisfaction.

Design Dealing with User's Needs

The meaning and functions of spatial requirements and structures differ greatly with regard to social class, ethnic groups, age, sex and so forth. Moreover, people's needs very greatly. In contrast, researchers have delineated some universal needs that are relevant for majority of users. Krupat (1987) has utilized the behavioural approach to urban design and identified six types of needs that are common to all people. These are:

1. *Security:* One of the most basic concerns of residents is feeling of safety and security in the places where one lives. This issue is relevant to all levels of design. People will not be much concerned with the satisfaction of higher order profession needs unless they perceive that the place they are residing is safe for themselves and their family members.

2. *Clarity*: The environment must be legible. Design that confuse and disorient residents in a large housing complex dissatisfies residents.

3. *Privacy:* People have strong desire to regulate social interaction with others. This issue is very relevant in large housing complexes where residents belonging to different age, sex and other demographic characteristics can get their desired quota of privacy in their home and neighbourhood.

4. *Social interaction*: By nature people are social. A number of studies have pointed out that both too little or too much social interaction are perceived as disturbing and that people exhibit strong desire for achieving optimal amount of social interaction with others. Furthermore, studies have described the importance of physical features of the designed environment in terms of its ability to orient people towards each other (sociopetal) or to pull people apart (sociofugal). Thus, the physical feature of the built environment, which provides optimal level of social stimulation, generates positive emotional states and residential satisfaction.

5. *Convenience:* The issue of convenience is related to the physical need of people. For instance, the connectivity of the residential environment with important day to day facilities like transport system, shopping, work place, school and other facilities are potent consideration which is related to various aspects of residential satisfaction.

6. *Identity:* The term identity refers to the relationship between people and environment. Usually, people display strong identification with the place they reside or work. People are emotionally, sentimentally or symbolically attached to the places they live for a prolonged period of time and any dislocation creates discontent and dissatisfaction (Fried, 1963).

In short, it is important for the architects to be aware of

the full range of human needs while planning various types of designed environment. The planners and architects should integrate both the aesthetic features and functional aspects of the designed environment. A designed environment that strictly emphasize on aesthetics ignoring the functional dimensions or vice versa is likely to be less satisfying for the users. In a similar vein, an overemphasis on functional dimensions often ignores cognitive, affective and social aspects as well as people's need for social stimulation. For instance, a fully functional design are efficient in controlling external environment conditions like heat, light and temperature and meet all building, housing and zonal codes. However, they are often judged as cold and sterile for it lacks complexity and variety.

Designed environment that is pleasant to eyes and is high on various dimensions of aesthetic may be problematic. It is often noted that purely aesthetics design sometimes neglect the more basic and instrumental need of users. For example, a beautiful building of a mental health centre located in Boston won design awards but was not appreciated by users and others in a long run because it lacked appropriate orientation. It was difficult for even frequent visitors to find correct passageways and staircases. Thus, aesthetically superior designed environment that ignores functional aspects and are not users friendly are confusing and problematic (Peake & Leonard, 1971; Weisman, 1979; Krupat, 1987).

While awareness of the range of human needs is important but it is also important to understand that not all needs are equally important for all groups at all time intervals. Rainwater (1966) has pointed out that the home and its surroundings have different meanings for different people. For instance, the urban poor, the traditional working class and the modern working class attach different types of thinking about their dwelling environment. Urban slum dwellers think about their home as a haven from fear. In contrast, the middle class people attach symbolic value or status to their home. In other words when defining or thinking about the home, it is important to realize that people belonging to different categories differ in terms of their needs, preferences and values. Thus what is best for one group of people might not be best for the other categories of people.

Design professionals and planners should use the knowledge of behavioural science or collaborate with social scientist to reduce the design problems. Another way to deal with the user-needs gap is to involve users in planning and design. A number of researchers have discussed the significance of user-generated design (Sommer, 1972). Community meetings may be conducted for getting users ideas directly before the planners. Other techniques such as role play and data collection jointly done by planners and social scientist may help design professionals to place themselves in users perspective (Burnette, Moore, & Simek, 1973). Good planning involves taking decisions that are based on systematic knowledge gathered from different perspectives that maximizes the salient needs of the users. Furthermore, post-occupancy evaluation research must be carried out to assist professionals with behavioural data responsive to people's needs. It seems that a combined knowledge incorporating physical aspect of design and behavioural aspects of users needs may be more effective.

Summary

Based on the studies conducted in a range of residential settings one can conclude that an interface between planners, architects and potential users is essential. It is observed that planners made decision on behalf of the potential users. This lack of consultation exposes the conflicts that can occur between the perceptions and interests of planners and the people involved. Thus, it is clear that participation in the design process of stakeholders is essential if the designed environment is to be effective. A lack of consultation is likely to lead to misinterpretation of the needs of residents. It is clear that when residents move to new dwelling where social networks are disrupted and if the physical features of the new dwelling environments fail to meet the social interaction and affiliation needs of the users various mental health related problems and residential dissatisfaction are likely. Sufficient empirical evidence exists that the architects are more concerned about the space utilization and aesthetics of the building and surrounding and are less aware about the psychological effects of the new environment. However, the role of various psychological factors including physical and functional proximity, location of amenities and services, homogeneity of neighbourhood, sense of community and so forth, are found to be helpful in enhancing residential satisfaction, positive emotional states and less mental problems. As Rohe (1985) suggests participation in the design process has effects on two main mediators of mental health, control and social support. Being involved provides a sense of control and empowerment and if done effectively allows some real control over the design process. Involvement also increases the interaction between residents, provides a sense of community and social support. Thus, designing residential environment the inputs from the users and their involvement in the entire process of design is needed together with adequate consideration of psychological factors may prove beneficial in enhancing the quality of life of the inhabitants and will be helpful in making the designed environment effective.

REFERENCES

Baum, A., Aiello, J., & Calesnick, L. (1978). Crowding and personal control: Social Density and the development of learned helplessness. *Journal of Personality and Social Psychology*, 36, 1000-1011.

Baum, A., & Paulus, P. (1987). Crowding. In D. Stokols and I. Altman (Eds.), *Handbook of Environmental Psychology* (Vol. 1, pp. 553-570), New York: John Wiley.

Baum, A., & Valins, S. (1977). *Architecture and Social Behavior: Psychological Studies in Social Density*. Hillsdale, N.J.: Erlbaum.

Bechtel, R.B. (1977). *Enclosing Behavior*. Stroudburg, PA: Dowden, Hutchinson & Ross.

Brolin, B. (1972). Experts planned Chandigarh but something has gone wrong. *Smithsonian*, 3, 56-62.

Burnette, C., Moore, G.T., & Simek, (1973). A role-oriented approach to problem Solving groups. In W. F. E. Preiser (Ed.), *Environmental Design Research* (Vol. 1). Stroudsburg, Pa: Dowden, Hutchinson & Ross.

Caplow, T., & Forman, R. (1950). Neighborhood interaction in a homogeneous community. *American Sociological Review*, 15, 357-366.

Carstens, D.Y. (1985). *Site Planning and Design for the Elderly: Issues, Guidelines and Alternatives*. New York: Van Nostrand Reinhold.

Caspi, A., Bolger, N., & Eckenrode, J. (1987). Linking persona and the context in the daily stress process. *Journal of Personality and Social Psychology*, 52, 184-195.

Cassidy, T. (1997). *Environmental Psychology: Behavior and Experience in Contact*. Sussex: Psychology Press Publishers.

Christensen, D., Carp, F., Cranz., & Wiley, J. (1982). Objective housing indicators as predictors of subjective evaluations of elderly residents. *Journal of Environmental Psychology*, 12, 225-236.

Cohen, S., & Syme, S.L. (Eds.). (1985). *Social Support and Health*. New York: Academic Press.

Ebbesen, E.P., Kjos, G.L., & Konecni, V.J. (1976). Spatial ecology: Its effects on choice of friends and enemies. *Journal of Experimental Social Psychology*, 9, 505-518.

Festinger, L., Schachter, S., & Back, K. (1950). *Social Pressures in Informal Groups*. New York: Harper & Row.

Fried, M. (1963). Grieving for a lost home. In L. J. Duhl (Ed.), *The Urban Condition*. New York: Simon & Schuster.

Galster, G. (1987). Identifying correlates of the dwelling satisfaction: An empirical Critique. *Environment and Behavior*, 19, 539-568.

Gans, H.J. (1967). *The Levittowners*. New York: Pantheon Books.

Gubrium, J.F. (1974). Victimization in old age. *Crime and Delinquency*, 20, 245-250.

Hourihan, K. (1984). Context-dependent models of residential satisfaction. *Environment and Behavior*, 16, 369-393.

Ittelson, W.H., Proshansky, H.M., Rivlin, L.G., & Winkel, G.M. (1974). *An Introduction to Environmental Psychology*. New York: Holt, Rinehart & Winston.

Krupat, E. (1987). *People in Cities: The Urban Environment and its Effects*. New York: Cambridge University Press.

Kuper, L. (1953). Blueprint for living together. In L. Kuper (Ed.), *Living in Towns* (pp. 1-202). London: Cresset Press.

Mawby, R. (1977). Defensible space: A theoretical and empirical appraisal. *Urban Studies*.14, 169-179.

Merry, S. (1981). Defensible space undefended: Social factors in criminal control through environmental design. *Urban Affairs Quarterly*, 16, 397-422.

Merton, R.K. (1948). The social psychology of housing. In W. Dennis (Ed.), *Current Trends in Social Psychology* (pp. 163-217). Pittsburgh, PA: University of Pittsburg Press.

Michelson, W. (1970). *Man and his urban environment: A sociological approach. Reading, MA: Addison-Wesley.*

Newman, O. (1972). *Defensible Space*. New York: Macmillan.

Newman, O., & Franck, K. A. (1980). *Factor Influencing Crime and Instability in Urban Housing Developments*. Washington, D.C. National Institute of Justice.

Paulus, P.B., & Nagar, D. (1987). Environmental influences on social interaction and Group development. In C. Hendrick (Ed.), *Group Processes and Intergroup Relations* (pp. 68-90), Beverly Hills: Sage.

Paulus, P.B., Nagar, D., & Camacho, M. (1991). Environmental and psychological Factors in reactions to apartments and mobile homes. *Journal of Environmental Psychology*,11, 143-161.

Paulus, P.B., Nagar, D., Larey, T., & Camacho, M. (1996). Environmental, lifestyle, and Psychological factors in the health and well-being of military families. *Journal of Applied Social Psychology*, 26, 2053-2075.

Peake, P., & Leonard, J.A. (1971). The use of heart rate as an index of stress in blind pedestrians. *Ergonomics*, 14, 189-204.

Perlman, J.E. (1976). Grassrooting the system. *Social Policy*, 7, 4-20.

Porteous, J.D. (1977). *Environment and Behavior; Planning and Everyday Urban Life*. Reading, mass: Addison-Wesley.

Proshansky, H.M. (1978). The city and self-identity. *Environment and Behavior*, 10, 147-169.

Rainwater, L. (1966). Fear and house-as-haven in the lower class. *Journal of American Institute of Planners*, 32, 23-31.

Rapoport, A. (1985). Thinking about home environment: A conceptual framework. In I. Altman and C. M. Werner (Eds.), *Home Environments* (pp. 255-286). New York: Plenum Press.

Rohe, W. (1985). Urban planning and mental health. *Prevention in Home Sciences*, 4, 79-110.

Sarason, S.B. (1974). *The Psychological Sense of Community: Prospects for a Community Psychology*. San Francisco: Jossey-Bass.

Schachter, S. (1959). *The Psychology of Affliction*. Stanford, CA: Stanford University Press.

Shaver, P., & Buhrmester, D. (1983). Loneliness, sex role orientation and group life: A Social needs perspective. In P.B. Paulus (Ed.), *Basic Group Processes* (pp. 259-288). New York: Springer-Verlag.

Stamps, A.E. (1994). All buildings great and small: Design review from high-rise to Houses. *Environment and Behavior*, 26, 402-420.

Sundeen, R.A., & Mathieu, J. (1976). The fear of crime and its consequences among Elderly in three urban communities. *Gerontologist*, 16, 211-219.

Stokols, D., & Shumaker, S.A. (1981). People in places: A transactional view of settings. In J.H. Harvey (Ed.), *Cognition, Social Behavior and the Environment* (pp. 441-488). Hillsdale, NJ: Lawrence Erlbaum.

Sundstrom, E., Bell, P.A., Busby, P.L., & Asmus, C. (1996). Environmental psychology, *Annual Review of Psychology*, 47, 485-512.

Taylor, R.B., Gottfredson, S.D., & Brower, S. (1981). Territorial cognitions and social Climate in urban neighborhoods. *Basic and Applied Social Psychology*, 2, 289-303.

Wallis, A.D. (1981). *Mobile Homes: A Psychological Case Study of Innovation in Housing*. Unpublished Ph.D. thesis, University of New York.

Whyte, W.H. (1956). *The Organization Man*. Garden City, NY: Doubleday.

Wandersman, A., Florin, P., Chavis, D., Rich, R., & Prestby, J. (1985). Getting together And getting things done. *Psychology Today*, pp. 64-71.

Weisman, G. (1979). *A Study in Architectural Legibility*. Unpublished doctoral dissertation, University of Michigan.

Unger, D.G., & Wandersman, A. (1985). The importance of neighbors: The social, cognitive and affective components of neighboring. *American Journal of Community Psychology*, 13, 139-169.

Zeisel, J. (1981). *Inquiry by Design: Tool for Environment-Behavior Research*. Montery, Calif: Brooks/Cole.

8

HUMAN RESPONSES TO PROTECT THE ENVIRONMENT

In its quest of conquering the environment man has ruthlessly plundered the ecosystem so much so that the quality of air, water and land is adversely affected causing hardship to mankind. The seriousness of environmental problems of technological industrial society have forced us to think about its consequences and to find out the ways by which the environment is protected from further destruction. Attempts are been made to find out the ways of preservation, conservation and sustainable development of ecology so that a proper balance between man and the nature is restored. In this chapter three environmental themes are discussed. In the first section attempt is been made to discuss and examine various intervention techniques, which may help in changing the behavioural pattern of the users so that environmentally destructive tendencies are arrested. In the second section, some major environmental movements carried out by people to protect the surrounding ecology in Indian context are discussed. Finally, two competing theoretical viewpoints with respect to addressing the environmental issues will be presented.

Changing the Environmentally Destructive Tendencies

The phenomenon of "Social trap" or "Common Dilemma" is helpful in explaining the reasons why people engage in environmentally destructive acts and behaviour. Breaking the 'Social trap' is difficult largely because of the often common thinking that everyone's business is no one's business. Most people are usually guided with this type of mindset. They hold strong conviction that they alone cannot do anything to save the outside environment that belongs to everyone. The novel entitled *The Tragedy of the Commons* written by Hardin (1968) offers an excellent analogy to explain many aspects of contemporary lifestyles. The story is centred on a shepherd who shares a common pasture known as "the commons" with other shepherds of the village. The land cannot be enlarged and all the shepherds use to utilize the same piece of land to graze their animals. Suppose you share the same pasture and the financial gains from sheep grazing are yours. To enhance your financial gains you decide to add one sheep for your flock. Yours decision to purchase one additional sheep and increase your financial gains is based on rationality and logic because you are convinced that by just adding one sheep will not present any pressure on the existing pasture. Your action of additional sheep purchase is based on a logical presumption that the cost of one additional sheep grazing will be very low. However, what will happen to the commons if every shepherd use the same logic and adds one sheep to their existing flock? The result will be gradual depletion of the commons and all shepherds whether they have or they have not added sheep would suffer. In these kinds of situations everyone is the looser and no one is going to be benefited. This story offers an excellent analogy with many aspects of contemporary lifestyles. The existing resources are overexploited to the extent that their sustainability for future use is under question. At the individual level we find ourselves amidst various types of such resource related decisions as one discussed earlier. In other words satisfaction of our immediate needs takes precedence over the long-term societal needs or not is a dilemma which most of us encounter in our day-to-day life. If all members of the society are able to understand that our short-term personal gains will be harmful both at the individual and societal levels in long run then everyone will be benefited. In contrast, if we fail to understand this logic than everyone will suffer. All the society members have to give up some portion of our freedom for the commons to survive. Thus, to enjoy the true essence of life harmonious relationship between man and the surrounding ecology is to be maintained.

In the preceding chapters a large number of studies conducted in the area of environment and behaviour focusing on the effects of physical environmental stressors like density, crowding, noise in a range of settings are discussed. These studies have dealt with examining the adverse effects of stressors on various behavioural, psychological and health related measures. It seems surprising that only very few attempts have been made to study the behavioural modification techniques which may prove to be helpful in preventing or minimizing the negative effects of these stressors or save the environment from further destruction. In other words more focused research is needed to study how positive affective states may be attained in response to environmental determinism rather than to study the damaging effects of environment on humans.

Environmental Psychologist should not only attend to pressing day-to-day problems faced by large number of people like shortage of fuel, air, water, noise and people pollution. Instead the studies should also focus on appropriate behavioural intervention strategies, which may be helpful in minimizing the adverse effects. Scholars working in physical science disciplines have emphasized on technologies like solar and nuclear energy as well as recycling the waste materials so as to reduce the demand of fuel shortage. Indeed the application of physical technologies with modification of human behaviour is needed to improve the surrounding ecological quality on a day-to-day basis. Evidence exist that technologies based on physical science knowledge, if used in isolation may solve problems pertaining to food, mobility, energy shortage but their side effects are innumerable. For instance, cotton mills have eased the clothing problem of masses but have contributed to health related problems like asthma, lung disease in masses residing in the vicinity of cotton industries. In a similar vein excessive use of packaging materials like plastic has not only caused litter problems but has also spoiled the soil quality to a considerable extent.

Similarly, overuse of transportation facility has increased mobility but has also caused problems of air pollution. It is possible that the side effects resulting from technological advancement may be minimized if proper behavioural interventions are planned. Various pollution control technologies are available to reduce the pollution. However, user's attitudinal and behavioural patterns has to be changed to deal with the adverse effects. For instance, catalyst converters, use of unleaded petrol, stopping the car engine at the busy intersection, periodic vehicle maintenance, driving the car at a constant speed, sharing the car with others, frequent use of public transport system may help in reducing air pollution and fuel consumption. It seems that behavioural change is essential if one intends to reap the long-term benefits of technological advancement. Granted the fact that the modern technologies has both positive and negative effects on human beings, one pertinent question relates to how technology is put to use? People in general are guided by conventional thinking. They suspect that changing behaviour to deal with the ill effects of scientific technology will adversely affect their quality of life. However, careful long-range analysis is required to examine the relationship between technology, behaviour and quality of life. It seems that behavioural change rarely affects the quality of life. For instance, if more and more people start using car or public transport, use cycle or walk if distance is short, the air pollution level will certainly drop and dependence on petrol or oil products will be minimized. However, these changes require change in behavioural perspective.

Social scientists usually adopt behavioural intervention techniques to deal with three categories of environmental problems: Environmental aesthetics (e.g. control of garbage in public areas), health related problems (e.g. noise, air pollution) and resource related problems (e.g. over consumption of non-renewable resources like petrol). Research studies conducted by Cone and Hayes (1980) focuses on promoting the environmentally protective behaviour (e.g. picking up litter/ garbage, recycling waste materials) and discouraging environmentally destructive behaviour (e.g. splitting tobacco, throwing plastic and cans in open spaces). The government and NGOs committed to environmental issues may think of various behavioural oriented strategies to deal with environmentally destructive tendencies and reinforce protective approach towards environmental protection. For instance, devising attractive remuneration packages to a large number of people who perform environmentally protective acts like picking up plastic bags, cans and other packaging materials from public places. In contrast, to discourage environmentally destructive behaviour high fines may be imposed. To solve some of the problems, which confront large section of society, it seems appropriate to discuss various behavioural oriented intervention strategies. A brief discussion of these techniques are presented below:

1. Environmental Education
2. Environmental Prompts and Cues
3. Reinforcement Strategies

Environmental Education

One of the most important objective of environmental education is to properly educate and convince people so that they become more sensitive and aware about the scope and nature of environmental problems and its adverse consequences. This approach may lead to formation of pro-environmental attitude and may consequently lead to behavioural change in masses. Proponents of educational inputs hold strong favourable opinion that information pertaining to environmental issues may generate awareness, which may subsequently lead to attitudinal and behavioural change (Left, 1974).

Educators frequently use different behavioural techniques and methods to cultivate pro-ecological attitudes with respect to dealing with environmental problems. The government is funding educational institutions and NGOs to conduct seminars, workshops and conferences so as to sensitize and generate awareness among people about the deleterious effects of environmental degradation. Mass media has proven to be very useful in disseminating useful information about the status of the environment in terms of air quality, water pollution, people pollution and so forth. Ways are suggested to reduce the impact of a range of environmental problems. To generate awareness courses on various aspects of environmental education and awareness are already introduced at the elementary, school and college levels.

Empirical studies are conducted to examine the effectiveness of educational programmes pertaining to various aspects of environment and its impact on changing the attitudes and behaviour. Mixed results are obtained with some scholars reporting enhanced awareness and positive behavioural change (Cohen, 1973; Allen, 1972) while others demonstrating either weak or no links between educational inputs and subsequent behavioural change (Cone & Hayes, 1980; Lingwood, 1971; Winston, 1974). For instance, Asch and Shore (1975) conducted a study on the impact on educational environment on school children for a period of two years. Their results indicated that relative to control group, children exposed to field trips to a natural area exhibited greater respect for wild life, plant life, soil and water. Based on the results obtained from this study one can infer that environmental education may be successful with young children who do not have well grounded habits. Presenting solution rather than increasing awareness may enhance the effectiveness of educational programs is noted by some scholars (Rankin, 1969). Some researchers have reported evidence that certain specific attitudes could be affected by environmental education. For instance, Olsen (1981) and Seligman *et al.* (1979) reported that people attitudes regarding energy conservation by lowering the thermostat is linked with health related problems while comfort is related to energy non conservation, and believing that energy crisis will have direct effect on oneself is positively linked with conservation behaviour. However, precise assessment of environmental education and its impact on attitude and behavioural change is needed to draw firm conclusion. It seems certain that even if the linkages between environmental education and consequent behavioural change are weak, education may serve a useful function in reinforcing pro-ecological behaviour and attitudes for at least those people who have commitment to contribute to environmental problems.

Environmental Prompts and Cues

The common environmental prompts are cues that convey or communicate simple written messages or verbal messages to the masses. These messages are presumed to facilitate pro-environmental behaviour and to discourage acts that are environmentally destructive. Some prompts are general while other is specific in nature. For example prompt displaying message like "Keep Delhi Clean and Green" is general in nature while prompt designed to communicate motor drivers to maintain a constant speed of 55 km per hour is specific in nature. In a similar vein some prompts signifies incentives (e.g. 10 paise incentive for collection of each plastic bags) while other types of prompts signifies disincentives for some specific behavioural tendencies (e.g. fine of Rs. 100 for wrong parking). In other words approach and avoidance prompts are used to encourage desired or discourages undesirable behaviour. For instance, in recent years the Election Commission of India has conveyed messages in the form of prompts cautioning that the writing on the walls in support of any party will result in high monetary fine. Thus, disincentives were attached by avoidance prompts with great degree of success. It seems that prompts moderate environmentally destructive behaviour by making certain social norms (e.g. prompts about specific vehicle fuel to be used, anti-litter slogans) either more salient or less salient in a given situation. Various studies have attempted to examine the impact of prompt on human behaviour. For instance, Haynes and Cone and Hayes (1978) found that prompts displayed in a newly constructed mini park indicated messages like "don't trample the grass" resulted in significant reduction in environmentally destructive lawn walking. A very similar trend of result was obtained by Baltes and Hayward (1974). In their study it was found that littering at a football game was reduced by 45 per cent by a written prompt.

A number of studies have revealed that people in general have the tendency to ignore prompts (Geller *et al.*, 1982). They observed that if the cost of adhering to the prompt is perceived as high as there are little chances of behavioural change. For instance, despite the prompt for pollution control to avoid using private cars during smog episode, no change in traffic flow was found in San Francisco. Various determinants of the effectiveness of prompts in bringing about change in attitudes and behaviours are identified:

1. Prompts with contingencies (rewards/punishments) increase the likelihood of pro-ecological behaviour.
2. Prompts, which are specific, but without contingencies are effective in situations when pro-ecological responses/behaviour are easy to emit. In other words, opportunities to perform the pro-ecological act are located in close proximity. For instance, pro-ecological actions are facilitated if attractive dustbins are kept near the displayed prompt stating, "please use me".
3. Wordings written in prompt that tend to create psychological reactance under certain conditions are reported to be ineffective. For instance, don't you dare to litter the park" is less effective than a anti-litter prompt requesting the users of the park "please, don't litter the park".
4. A prompt of a moderate size and salience is more effective than a prompt, which is of large size and high salience. The very large and salient prompt creates reactance in users, which minimizes the impact of the message.
5. A human model displaying some specific behaviour may also serve as a prompt. The models provide useful information about the expected behaviour (e.g. anti-litter slogans) and also about the positive and negative consequences to the masses.
6. The physical presence of an object like trash case may serve as effective prompt and is likely to prohibit environmentally destructive behaviour.

Reinforcement Strategies

A range of reinforcement techniques can be used to modify the environmentally destructive behaviour or improving the quality of the environment. The common reinforcement strategies are positive, negative, punishment and feedback. These reinforcement techniques are consequence strategies directed to change/modify the attitude and behaviour of users. A brief discussion of some common reinforcement techniques are given below:

Positive Reinforcement

Individuals are given positive reinforcement for performing environmentally friendly and constructive behaviour. The underlying premise is that reinforcement serves to increase or strengthen the probability of desirable and favourable behaviour. The positive reinforcement strategies can be grouped into two categories. One group consists of techniques, which are aimed to improve the already disturbed environment, while the other group of studies focuses on preventing negative or environmentally destructive behaviour from occurring at the first place. Administration of positive reinforcement is easy on the one hand and cost effective on the other. The reinforcement may range from tangible financial rewards to non-monetary benefits. Sometimes positive reinforcement is response contingent and sometimes they are response non-contingent (e.g. achievement of particular outcome). Ingram and Geller (1975) provided response non-contingent reinforcement in the form of giving raffle tickets for bringing waste papers to a recycling bin while giving some monetary benefits for cleaning the environmental mess is an example of outcome contingent reinforcement. The response contingent reinforcement is more effective relative to outcome contingent reinforcement because the former provides information about the desirable pro-ecological behaviour (Geller *et al.*, 1982).

Schmidt and Ulrich (1969) conducted research focusing

on noise elimination on elementary and secondary school students. The researchers observed that the average noise level was 52 dB. However, when the students were given an impression that if the noise level is reduced to 42 dB an incentive in the form of extra gymnasium will be given. Favourable results in the form of noise reduction were obtained. Wilson and Hopkins (1973) obtained more or less similar trend of results in their study on noise elimination. Thus, on the basis of results obtained one can infer positive reinforcement may help in improving and restoring the disturbed environmental conditions. Some studies have attempted to address issues related to prevent possible negative effects on environment. The researchers focus on reinforcing people for exhibiting positive environmental behaviour so as to prevent the anticipating negative effects. For instance, to promote energy conservation both financial and non-financial incentives have been used successfully. The use of non-monetary rewards for commuters who either use public transport facility or carry three or more passengers in a car. The non-monetary benefits were given in a form of offering the car drivers to drive their cars in a special lane with fast traffic flow. Use of public transport system may reduce the tension of the car driver to drive their vehicle in busy traffic or finding a parking place for the car. While supportive evidence is available, however, the effectiveness of such programmes in preventing negative environmental effects is not systematically researched.

Negative Reinforcement and Punishment

These methods have been used less frequently as compared to positive reinforcement to control or modify environmentally destructive behaviour. The negative reinforcement may involve avoidance of unpleasant aspects of the situation which punishment is operationalized in terms of adding aversive elements to it. In a number of countries and cities located in Europe and United States some common uses of punishment are collecting fines from people who happen to throw trash in open areas of roadside. Furthermore, the companies who are responsible for crossing the limits of pollution are fined. In the Indian context similar measures should be taken to prevent people from splitting or throwing waste products in streets or in common areas of public use. Similarly, industries exceeding the pollution limits should be financially penalized. In some big cities the government is attempting to use negative reinforcement and punishment to those people whose vehicles fails to adhere to pollution norms. In a number of countries parking charges are taken from the vehicle owners. The rate of parking varies from the peak hours to the steady hours. In many big cities in India the parking charges are taken from the vehicle owners who park their vehicles in busy market places or railway stations. The norms regarding parking are to be clearly laid out and violation of the norms may be strictly dealt with associated fines. In a similar vein, studies have reported that aircrafts exceeding the prescribed noise levels may be prevented from landing at convenient airports (Cones and Hayes, 1980). The noise pollution related guidelines might caution the aircraft manufactures to reduce noise levels of aircrafts so as to prevent prohibitions. Furthermore, punishment or negative reinforcement in the form of cutting down corporate tax benefits for companies polluting air, water may be used to combat environmental problems.

Feedback

This technique is used to communicate to the people how they are responding to various aspects of the environment. Often these techniques are used in combination with other strategies like environmental prompts. Estimates of community litter may be given at periodic intervals in the newspapers or television, which may serve as a feedback to people in general.

Integrated Approaches for Eliminating Environmentally Destructive Actions

While each of the techniques discussed in the preceding section has its own benefits, it seems reasonable to expect that a combined strategy using various individual methods may prove to be more effective in dealing with specific qualities of environmental problems and environmentally destructive behaviour. Meyers, Artz and Craighead (1976) in their study attempted to deal with "noise nemesis" utilized elements of both reinforcement and environmental education techniques. The reinforcement component consisted of monetary rewards while for the feedback a sound of doorbell was used when serious noise transgression occurred. Finally, the environmental education component included inputs about how to deal with the noise. The results of this study revealed that the combined strategy was effective with striking decrease in noise transgression from 345 per day to 145 per day. The relative impact of various individual methods used in combined strategies indicate that reinforcement strategies are strongest followed with prompts while the impact of environmental education is reported to be weakest (Cones & Hayes, 1980; Clark, Hendee & Baurgess, 1972). The study conducted by Clark *et al.* (1972) investigates the relative impact of several antilittering strategy in a movie theatre. In the environmental education approach the Walt Disney antilittering cartoon was displayed while prompt included distribution of litterbags and instructions or adding extra trash cans. Finally, reinforcement techniques provided money or movie tickets to the subjects. In each condition the proportion of the total litter deposited was recorded. The results indicated that reinforcement in the form of money or free movie tickets proved to be most effective. The use of prompts was moderately effective while the environmental education method was least effective in dealing with littering behaviour. However, before deciding on the strategies with a particular environmental problem it seems essential to weigh the relative cost/benefit of a particular programme. The reinforcement effects are most pronounced but are most expensive to administer relative to other strategies. In contrast, the prompts are relatively inexpensive relative to other strategies but are less expensive ways in dealing with undesirable environmental behaviours. Thus, it seems that a combined approach is most effective as compared to individual approaches in dealing with environmental problems and issues.

Approaches to Deal with Specific Environmental Problems

In this section attempt will be made to discuss some frequently used techniques for combating with environmentally destructive behaviour and actions. Applications of some specific approaches will be discussed to deal with minor or major environmental problems that are of concern to a large section of society.

Garbage Problem and Littering Behaviour

In almost all the Indian cities and rural areas trash, garbage and waste products are commonly visible in public places. The problem of garbage and littering not only gives a shabby appearance to the surrounding but it dangerously affects the health and well-being of masses. With some exception chewing beatles and splitting on roads, streets are so common that it has become an inescapable part of community lifestyle in most of the places in India. In order to combat the problem of splitting of beatles and tobacco in some major railway stations selling of tobacco products are strictly prohibited. Moreover, cans are kept in close proximity so that garbage and waste products are not thrown on railway tracks. It seems that to some extent these attempts by the railway departments have proved helpful in reducing the amount of litter thrown by users in the open spaces. However, comprehensive strategies like reinforcement, sanctions and prompts may be used to motivate and prevent careless people from littering.

Energy Conservation: Saving Energy at Home

Researchers focusing on energy and fuel conservation have observed that approximately 20 per cent of the energy is consumed in residential settings in United States and much of this consumption can be treated as wasteful (Cone and Hayes, 1980). The studies conducted by Socolow (1978) reveal that some simple physical and behavioural changes may reduce 50 per cent of the energy consumed in a typical American home. In India it is estimated that 30 per cent of the energy is consumed in residential areas. A good proportion of energy consumed in cooking can be saved through good cooking habits and with some common precautions. Some of the tips for household cooking are as follows:

- Pressure cooker cooks faster and saves 20 to 46 per cent of fuel.
- Use optimal water while cooking. Surplus water consumes 65 per cent additional fuel.
- Cooking time is reduced provided the dal, rice and beans are properly soaked.
- While cooking always cover the vessel or utensil to avoid loss of heat in the atmosphere.
- Regularly clean the burner and replace wicks of kerosene stoves.
- Develop the habit of keeping the frozen food out of the fridge for some period before putting them on the gas or stove.

Thus, it seems reasonable to expect that change in habits and attitudes at the individual level may contribute effectively in dealing with energy crisis to a large extent. Propagation of residential energy conservation must be carried out through environmental education. Brochures, leaflets and pamphlets may be prepared by the energy department and may be distributed to the public. The involvement of NGOs may further help in the dissemination of information about optimal energy use and its effective consumption. Leaflets may be kept inside the newspaper for wider circulation. In addition, people in general should be educated about the utility of energy conservation and tips about switching off the lights and fans when occupants are not using the area may be propagated at periodic intervals to the masses.

Energy Conservation

Concerted conservation efforts are required for optimal utilization of energy. *The Times of India* dated January 2000 New Delhi in its energy conservation supplement has presented a comprehensive package to deal with energy shortage at the industrial sector level as well as at the behavioural levels. A recent statistics shows that for importing petroleum products India spends Rs. 35,000 crores every year. India's average energy shortage is around 7 to 11 per cent. Thus, Rs. 26 crores is required to meet the shortfall. There are various common tips if followed at the individual level may help conservation of fuel to a large extent. The industrial sector particularly dealing in thermal, furnaces, boilers and electrical areas should adhere to periodic energy audits, proper combustion, observance of colour of smoke, air pressure, fuel losses, proper selection of pumps and motors. These steps are found to be effective way to deal with energy over consumption. An attempt is being made to review some precautions and tips which if used by masses in their individual capacity to combat energy conservation.

Fuel Conservation for Transportation

In recent years phenomenal increase both in the private and public transportation is visible. According to Everett (1977) about 30 to 40 per cent of all the fuel put to use is exclusively for transportation purpose. Not only do transportation contributes to air and noise pollution but its overuse and inefficient use significantly contributes to fuel wastage is amply noted in metropolitan and big cities. The report given by the Federal Energy Administration (1977) reveal that if each car in the United States were driven 10 miles per week less in United States, five percentage of the annual demand for gasoline would be reduced. In one study Foxx and Hake (1977) provided users of the vehicles various incentives (e.g. cash, free tour) to reduce the number of miles they drove in their car. The results of their study revealed that the participants who were provided with incentives drove their vehicles 20 per cent less in terms of the miles covered relative to the control group who drove their vehicles 5 per cent more. In a subsequent study Hake and Foxx (1978) found evidence that when a leader (e.g. teacher, supervisor) was the participant in the experimental group, the energy saving in the group was enhanced. Furthermore, evidence exists that competition between teams and other incentives like lottery tickets instead of cash payment also proved to be an added factor and motivator in mileage reduction programme (Reichel & Geller, 1980). Some studies have reported evidence that fuel rationing facilitates energy conservation (Cass & Edney, 1978). The results of the study on fuel rationing are best explained within the perspective of "commons dilemmas". It seems reasonable to expect that when "commons" are divided, people may use their portion more prudently and carefully. The results obtained on fuel rationing are of some value and significance. However, strong political will and conviction is required to put the fuel rationing concept practically into force at all levels to conserve the fuel resource. Another effective way to optimize fuel energy in transportation sector is to increase mileage per litre. The government may enforce laws or provide incentives to promote the manufacturers to design cars, which are more fuel-efficient.

A recent advertisement appeared in the *Hindustan Times* on January 17 to 31 on the eve of oil conservation fortnight attempt to educate the masses. It stated that "To save petrol you don't have to change your car, you have just to change your driving habits." Some of the educative tips for the motorist are as follows:

- Keep engine tuned and save upto 6 per cent on petrol consumption.
- Drive your vehicle in correct gear. Incorrect gear shifting can lead to as much as 20 per cent increase in fuel consumption.
- Maintain correct tyre pressure. Low tyre pressure can cause 5 to 10 per cent excess fuel consumption.
- Do not drive your vehicle with clutch pressed as it leads to fuel loss.
- Turn off the engine if you stop for more than two minutes and save upto 20 per cent petrol. Idling causes fuel wastage.
- Drive two wheelers at a constant speed of 35 to 45 km per hour.

- Drive four wheelers at a constant speed of 45 to 55 km per hour.
- Keep wheels properly aligned as improper alignment results in fuel wastage.

In India Railways is the most efficient transport system. To combat the menace of environmental degradation and in preserving precious oil reserves Northern Railway has taken numerous Eco-Friendly steps (*Hindustan Times,* New Delhi, 2001):

- Tree plantation all along the railway tracks.
- Use of concrete sleepers in place of wooden sleepers to avoid felling of trees.
- Introduction of synthetic material in place of wood in construction of the building.
- Enhanced axial load for longer train lengths resulting in clearance of a larger numbers within the existing infrastructure capacity.
- Gradual introduction of new technology coaches with low level of noise pollution.
- Regular disposals of scrap to maintain clean surroundings.

Environmental Movements

The people's movement addressing vital environmental issues pertaining to restoring, sustaining and maintaining the man-environment delicate relationship was initiated in the West. But over the last couple of decades the environmental movement has taken deep roots in many of the developing and third world countries. In India, individuals and voluntary groups have taken several bold initiatives to protect their surrounding ecology from destruction. For instance, the Chipko Andolan (movement) initiated by a group of villagers aimed at saving the ecology which was badly affected by exploitation of forest resources for commercial purposes. Adopting the method of non-violence this movement gained considerable popularity and success. The Chipko movement (Hugging trees from being cut) triggered Appiko movement in Karnataks. In the state of Kerala, the Silent Valley movement gained wide popularity and media coverage. This movement was focused on protecting the trees, destruction of forests and other allied activities like construction of big dams. The collective proactive environmental protection response from the local community was in most cases, supported by NGOs, environmental activists, social scientists and social workers received considerable media coverage, popularity and acceptance. Another visible movements, for examples, are the Fisherfolk's movement in Goa, Bodhgaya Land Struggle in Bihar, Mitti Bachao Andolan in M.P. These environmental movements adopted the Gandhian philosophy of non-violence like fasting unto death, *dharnas,* mass protest leading eventually to imprisonment by police. While it is difficult to cover all the people initiated endeavours to protect the ecology from destruction, a modest attempt is being made by presenting two detailed case studies of some highly visible environment movements initiated by local community with profound dedication and concerted effort to protect their surrounding ecology from destruction.

The Chipko Andolan

The connotation "Chipko" stands for "hugging" and "Andolan" stands for "movement". In other words the Chipko movement initiated by local people residing is the Himalayan region was aimed to protect trees by hugging them and saving them from being cut by contractors. Incidentally, the forest policy laid down in 1951, advocates that at least one-third of the country's land should be under the forest cover. Unfortunately, in India even half of the required area is not covered by forest. Growing urban demands and industrialization has hurt the forest badly so much so that we are losing forests at the rate of 1.3 mha every year (The state of India's environment, CSE, 1984-85). In the name of development destruction of forest continues unabated. Evidence exists that forest land is been used for various so called developmental and welfare projects like construction of dams, rivers and valley projects, industries, roads and agriculture. Forest vegetation is to be protected at the war footing before the situation reaches alarming situation. People living in the villages are the worst victim because their entire livelihood and lifestyle is dependent on the forest resources. According to Bhatt (1987) in the Himalayan areas of UP, forests have receded so far away from villages that women have to spend their entire day to reach the forest. They carry a load of 25 to 30 kg and walk over the difficult terrain to satisfy the basic primary need of their family.

Forest located in the Himalayan region keeps the atmospheric circulation in balance over the large part of Indian sub-continent. However, the result of large scale cutting of trees and heavy destruction of forests in the 60s not only degraded the natural environment but the daily lifestyle of the people was also seriously affected. Furthermore, in 1970s heavy damage was evidenced when the flood in the Alakananda River swept away bridges, footbridges, buses, villages, houses and acres of crops. The exploitation of forests only yielded an income of Rs. 10 lakh against the damage of several million of rupees (Bhatt, 1985). Despite this, without giving any thought the exploitation of forests for commercial use continued. Against the forest department, the people of Chamoli district made a series of protests between 1973-77. Following the Chipko Andolan in Reni a committee of experts agreed that the issues rose by the local people and villagers were scientifically correct and forest science was wrong. Chipko Andolan (the movement to hug trees to save them) caught the attention of the environmentalists and government and attempts were made to conserve the forests.

Six Points Memorandum of Chipko

To conserve the forests keeping in mind the needs, aspirations and expectations of the people living near the forests and with needs of the forests themselves a six point memorandum given to the government by the Chipko activists is as follows:

1. In a particular geographical situation, only specific trees and vegetation should be grown, so that appropriate needs of fertilizer, soil, water and energy can be met. Until the forest is not able to take care of its own survival, the support of human beings is needed for its protection. A detailed analysis of the social and economic conditions of the people living in the forest areas is being carried out. It should be understood that only people living near the forest could establish a practical and harmonious relationship with the forest and that the traditional relationship between local people with the forest should be seen and acknowledged.
2. The areas affected by the landslides and soil erosion or where forest areas are crucial for conservation of water resources should be identified and reserved. An old women involved in the Chipko movement made a remark that "The terror of tiger has gone. But it has been replaced by terror of landslide." The landslides destroy trees and adversely affect the fields.
3. The minimum needs of people living near the forest, who have been customarily using them for their survival, should be established and the administration should be such that they exercise these rights easily.
4. The contractor system should be completely stopped in forest conservation, development and exploitation. The people living near the forest areas should be organized to undertake all such activities.
5. Trees, which are useful to the local people, should be planted near the villages so that the people do not need to go into the reserved areas.
6. Village industries should be set up, based on minor forest produce so that the local population can get jobs and migration from villages can be stopped.

People's Movement for Afforestation

Keeping in mind the need for forests and wasteland development Prime Minister Rajiv Gandhi in 1985 spelled out the environmental, social and economic dangers resulting from the decline of forest cover. The establishment of National Wastelands Development Board was announced to meet the afforesting target of 5-mha wasteland with fuel and fodder trees. The Prime Minister announced that people's movement for afforestation is started.

The success rate of government programmes of tree plantation was very low relative to the trees planted by people's will. Evidence of people opposing the official afforestation and forest conservation was seen largely because the programmes were thrusted upon the people. The development and conservation of forests is not possible by making laws and issuing prohibitory orders. The government's policies, which allow the sales of forest produce in international and domestic markets at competitive price isolates local people who feel that their rights, are not taken care of. The Chipko Andolan believes that for any programme to be successful people aspiration's should be taken into account because these local people seem to identify themselves with the forests on which their lifestyles is heavily dependent. Furthermore, the government should not ignore the economic interests of the local people.

Chipko believes that unless cheep and alternative means for cooking is found people will continue cutting firewoods. Gobar gas plants and micro-hydel units could be set up to meet the energy crisis. However, all these alternatives are expensive relative to afforestation, which is the simplest. Besides solving energy crisis afforestation may be helpful in maintaining ecological balance and control the ravages of soil erosion, landslides and floods.

Women and Forest Conservation

Women in Reni village, which is located in the extremely sensitive area of the Alaknanda catchment, has called their forest their mother's home (*mayaka*). To save their forests in 1974 the lumberjacks: "This forest is our mother's home. If the forest is cut, the soil will be washed away. Landslides and soil erosion will bring floods, which will destroy our fields and homes, our water sources will dry up, and all the benefits which we get from the forests will be finished." Similarly, women from adjourning villages gathered together against the district administration to save their oak trees.

The Chipko Andolan made the forest administration realize the mistake of ignoring the women. An agitation launched successfully in 1980 by the women of Dungri-Paitoli village questioned the district authorities as to why their opinions were not seeked before auctioning the forest? The Panch and all male villagers unanimously surrendered their Oak Forest to the government so that the area was converted into a potato farm. However, the intervention of local women averted the auction. A series of similar Andolans launched under the broad banner of Chipko were organized and committed women villagers residing in the Chamoli district or near its vicinity made the andolan successful resulting in the saving the forests. Thus, it seems that any integrated programmes on forest conservation and management will only be successful if local community; specifically women whose lifestyle is dependent upon the forest are involved in policy formulation and implementation.

Experiences of Dasholi Gram Swarajya Mandal (DGSM)

The mother organization of the Chipko Andolan DGSM has taken several initiatives in the broad areas of environmental conservation and protection. Some of the significant work undertaken by DGSM is as follows:

1. Created awareness among the villagers about the importance of forests and seeking their active participation in afforestation and conservation.
2. A series of eco-development camps are organized in Central Himalayas and Alaklnanda region where problems pertaining to landslides and soil erosion are serious. Awareness is created among the villagers so that they realize the importance of forests in relation to their satisfaction of economic and social needs.
3. Every monsoon and winter several camps are

organized where people participate in large number. Local songs dealing with forest protection are played; educational discussions are held to address local problems and their solutions. People participating in the camps voluntarily donate their labour and plant trees. During tree plantation drums and various local instruments are played. To meet the local needs unutilized barren land are used to plant trees that are capable of meeting the local needs. The camps last for five to ten days where the atmosphere resembles a site of festival. Women from adjourning villages also participate in the camps and sharing of experiences facilitates learning.

4. DGSM sometimes provides technical and partial financial support.
5. DGSM is involved in promoting agro forestry as a mean of improving soil and water conservation along the Alakananda River and its tributaries.
6. The organization has successfully encouraged the villagers living near watersheds to protect their fields with stonewalls.
7. Soil erosion around the fields has also been prevented. Villagers who previously had to go miles to obtain grass for their animals are now getting grass near their fields.

The reason for the people's movement to be successful lies in their commitment and participation in ecological issues that are of immense concern to them and their families. People are ready to participate in all those developmental activities where they are requested to share their expert views in an atmosphere of trust. The government should understand that forest conservation and protection scheme if fails, to involve the local community is not likely to be successful in the long run because implementation of any scheme requires whole hearted support from local people. The relationship between people and the forest should be appreciated and honored. The local community residing in the forest areas for generations seems to identify themselves with the forest. They not only know the richness and variety of the forest but they are also aware of the severe consequences of deforestation. The government should understand the needs and aspirations of the local people particularly women. Micro level planning at the village level, involving cooperation, involvement of the local community is required to ensure conservation, protection and sustainability of the forests. The government should give control of forests in the hands of village communities or voluntary agencies. However, the government should provide support and assistance to the local villagers from time to time. A confidence building exercise with the spirit of equal partnership between community and government is needed to restore a harmonious balance between man and nature.

The Narmada Andalon

The Narmada river project began in 1979 is expected to be complete in 2025. The government proposes to build 3,200 big, medium and small dams along 1,200 km of Narmada river. The country's largest river valley project is going to benefit people from Madhya Pradesh, Rajasthan and Gujarat. While the Narmada Project is likely to displace 4,600 families but it is likely to provide drinking water for 8200 villagers, 135 towns and irrigate 1.91 million hectares of land (Dharchaudhuri, 2003). The Western model of development fascinated the politicians and others who started visualizing the huge river was letting its many million litres of waters go waste into the sea and that the water can be more productively used. The environmentalists believe that the construction of dams will damage the fragile ecology of the region and displace the inhabitants in very large numbers.

The Narmada passes between M.P. and Gujarat. People from both the states are continuously facing terrific shortage of safe drinking water. For instance, even before the closure of monsoon in many regions of Saurashtra water scarcity was visible with people getting water once in three days for twenty minutes (Buch, 2001). In the State capital of Madhya Pradesh,

Bhopal acute shortage of water is seen with water supply severely disrupted in many parts of Bhopal. The interstate rivalry and conflict started over the issue as to who should build a higher dam. Even the late Prime Minister Indira Gandhi was not able to settle the dispute and tribunal was set up to resolve the conflict regarding sharing of the water, power and investment involved. The World Bank agreed to fund the project. From 1979-83 works on the Narmada project progressed unabated and no significant queries were raised. Although the tribunal's opinion was technical one, it unfortunately failed to analyze the environmental impact.

The NBA strongly feels that government is not providing all the information to the public. Narmada Project is still in the news largely because the issues pertaining to good governance is conspicuously absent. The government has failed to provide all the relevant details about the decentralized water management system to the people concerned. Some of the questions raised by people are cultural in nature. These questions are related to saving trees in the dam region, disruption of downstream fisheries, submerging the forest farmland, making the water saline along the canals, the possibility of seismic faults and so forth. The people who are affected, the social workers and environmental activists are all consistently raising queries and allegations on various issues pertaining to how much land would be submerged, how many people will be uprooted, what will be the status of persons who loose land, what the government is doing to rehabilitate those families who are uprooted, who would be the benefited and so forth. In US all relevant information is provided to the people and dam fighters are invited to attend the conference. In many cases the proposals of dam construction are scrapped because the people were able to provide a socially and environmentally better alternative of water management. (Misra, 1987).

As has been reported in the *Hindustan Times* for the last twenty-three days four activists of the Narmada Bachao Andolan are observing hunger strike in Bhopal. Adopting the principle of non-violence these activists fasted two days longer than Gandhiji did on any of his fasts during the freedom struggle (Roy, 2002). The government of M.P. has forcible evicted more than thousands of Adivasi families to make their way for the Mann dam. The dam has been built. Adopting Ahimsa the only just war is being fought on a pavement in Bhopal still continues (Rao, 2002). The NBA activists wants that approximately 6000 displaced people must be settled before the reservoir fills up in monsoon and submerges their villages. The affected people have even identified the land that is available and could be allotted to them yet the government refuses. Before the heavy police force reached the dharna venue to forcibly break the agitation all the four activists holding fast for 23 days and were gradually loosing weight slipped away (HT, 12th June 2002). The CM stated that the state government is highly sensitive about the families affected by the Narmada Valley Development Project and has promised to provide relief according to the rehabilitation policy to all the affected persons. However, the activists feel that justice is not been given. Huge tracts of land is available for corporate agriculture but in the name of development the government is setting off another cycle of uprootment and impoverishment of displaced poor adivasis.

Dams are just a part, and a symbol of the developmental process. The powerful lobby in the society is making use of the natural resources at the expense of the poor and weak people who are thrown away, dislocated and uprooted from the places where they used to stay with their community members for generations. In October 2000, after the Supreme Court of India demolished the 'no dam' stance, the NBA activists including Medha Patker and common people have shifted their focus to relief and rehabilitation. The rehabilitation and relief packages outlined by NBA includes allocation of two hectares of land, a plot for housing, grant for ploughing fields, livestock, roofing, insurance, transport as well as community benefits such as schools, health centres, village ponds, well, parks, dispensaries and much more (Dharchaudhuri, 2003). However, actual participation between government and people is lacking. Nevertheless, the people's awareness regarding environmental issues and their commitment to deal with environmental concerns has certainly created substantial impact on policy and decision makers.

Theoretical Perspectives on Environmental Crisis

Two competing theoretical perspective—Deep ecology and Gaia framework will be presented to provide the reader with stimulation about environmental problems and related issues.

Deep Ecology

Worldwide people are becoming more and more aware about the unpleasant consequences resulting from the disturbed man-environmental relationship. Depletion of natural resources, depletion of Ozone layer in the atmosphere, massive deforestation, greenhouse effect, air, water and human pollution is at increase. Deep Ecology provides a solution to the environmental crisis advocating a radical shift in the way people relate with the environment.

The emergence of Deep Ecology movement was more or less spontaneously during the Ecological Revolution of the 1960s. This Deep Ecological movement has been characterized philosophically by a move from anthropocentrism to ecocentrism. However, Naese (1972) is credited to have coined the term "Deep Ecology"(Sessions, 1995). The notion of Deep Ecology emphasizes that we should think clearly and deeply about the ecological issues. Its main concern has been to bring about a major paradigm shift—a shift in perception, values, and lifestyles—as a basis for redirecting the ecologically destructive path of modern industrial growth societies.

Historical and Inspirational Roots of Deep Ecology

The birth of the Deep Ecology movement paralleled the rise to public prominence of the science of ecology and the "ecological perspective" as popularized by Aldo Leopold's ecocentric "land ethics" and from Rachel Carson, Dave Brower, Paul Ehrlich and other biologists, field ecologists and conservation organization leaders concerned with the rapid widening ecological crisis. These biologists and environmental activists were convinced that the dominant anthropocentric orientation of western civilization was seriously misguided and inadequate to deal with the crisis.

The Ecological Revolution of the 1960s is inspired from the publication of Carson's Silent spring in 1962. She questioned the direction and goals of modern technology as well as about the human health related adverse consequences resulting from the indiscriminate use of pesticides. Furthermore, she questioned the human competence and "right" to dominate and manage the Earth and pointed out that "control of nature" is a misleading phrase conceived out of arrogance. Finally, she advocated that the nature exists for the convenience of man. Historian Lynn White (1967) argues that environmental crisis is fundamentally the crisis of the West's anthropocentric philosophical and religious orientations and values. Christianity had desacralized nature, encouraged its exploitation, and promoted an anthropocentric worldview in which humans are superior to, and in charge of, the forces of nature. White claimed that modern science and technology are permeated with Christians arrogance towards nature. Marxism and other so-called "post-Christian" ideologies in the West also promote the same exploitive attitude towards nature. Finally, the views of Saint Francis who preached "the equality of all creatures" may be helpful in reforming Christianity ecologically.

Concept of Deep Ecology

The adjective 'deep' stresses on raising the type of deeper questions pertaining to Why and How about ecological issues. The concept of Ecosophy used by Naess deals with the tremendous expansion of our ecological thinking. It goes far beyond the narrowly phrased questions addressed by scientists. The term "sophy" is derived from the Greek language, which may be equated with ethics, rules, and practice. Ecosophy or deep ecology involves a shift from science to wisdom.

Deep versus Shallow Ecology

Eminent philosopher Naess (1995) has articulated the distinction between deep and shallow ecology, which is now accepted as an established terminology (Capra, 1995). The concept of shallow ecology is primarily anthropocentric. It views humans as above or outside of the nature, as the source of all values, and ascribes only instrumental or use value of the nature. In contrast, deep ecology believes that humans and the natural environment are inseparably interlocked and cannot be separated. It does not see the world as a collection of isolated objects but rather as a network of phenomena that are fundamentally interconnected and interdependent. Deep ecology recognizes the intrinsic values of all living beings and views humans as just one particular strand in the web of life.

In the last several decades massive disruptions of ecological processes and atmospheric conditions is evident. All forms of life in the planet Earth are threatened. It is believed that ecological problems have its roots in various variants of expansionary industrialism. Deep ecology provides a perspective to deal with environmental issues that are of concern to mankind. Deep Ecology as a theory as distinct from Deep Ecology as a social movement has become a benchmark in defining varieties of environmental philosophies (Fox, 1986).

General Views about Deep Ecology

The general views that supporters of Deep Ecology hold in common are discussed by McLaughlin (1995) and are subsumed in eight important points:

1. The well-being and flourishing of human and non-human life on earth have value in themselves. These values are independent of the usefulness of the non-human world for human purposes. This proposition is a rejection of anthropocentrism. It is an assertion that human and non-human life should flourish. Human being should take care of nature and should display a sense of relatedness to the nature.

2. Richness and diversity of life forms contribute to the realization of these values and are also values in themselves. The so-called simple, lower or primitive species of plants and animals contribute to the richness and diversity of life. Valuing diversity means freeing large areas of earth from domination by industrial economy and culture. This proposition stresses on preserving indigenous peoples as well as protecting wilderness.
3. Humans have no right to reduce this richness and diversity except to satisfy vital needs.
4. The flourishing of human life and cultures is compatible with a substantial decrease in human population. The flourishing of non-human life requires a smaller human population. The logic behind this proposition is to caution human beings that they have already jostled many species out of existence and many more are on the verge of extinction. Recent projections by the United Nations indicate that current trends in population growth will involve about 80 per cent of current nature reserves to human use (Sadik, 1992). This would drastically accelerate the already alarming trends towards the extinction of myriad species of life (Wilson, 1992).

   The continuing increase in human population causes many maladies including enhancing life of suffering to people. The Deep Ecology movement gives priority to human population as a problem and stresses the need to curb the population so that a decent life for all is ensured. In this regard joint effort is needed between Deep Ecologists and Ecofemininists. The problem of coerced motherhood leading to coerce pregnancy is particularly visible in poor countries where population growth is high. A reversal of this trend is needed. Women should be given the right to choose the number of children they would bear, rights to choose sexual partners and manage fertility in safe ways and should have access to safe abortions.

5. Present human interference with the non-human world is excessive, and the situation is rapidly worsening: The slogan of non-interference does not imply that humans should not modify some ecosystem, as do other species. The question is regarding the nature and extent of interference. The per capita destruction of wild forest and wild ecosystem has been excessive in rich countries. Human interference with the nature is rapidly at increase. Two major type of interference is as follow: (a) Destruction of existing areas of wilderness (age-old forests). The principle is to understand the continuation of biological history and creating large wilderness areas for plants and animals. (b) Technological disruption is gradually altering the natural cycles. For example, agricultural practices involving large-scale monocropping create enhanced needs for insecticides, pesticides and fertilizer. Multicropping, integrated pest management and organic farming techniques are found to interfere less with natural cycles and contribute to improve the soil quality.
6. Policies must be changed. These policies affect basic economic, technological, and ideological structures. The resulting state of affairs will be deeply different from the present: Attempts are been made to create adequate models for change. Now a days "sustainability" has become a slogan. Even those who believe in anthropocentric perspective are thinking how economics should be restructured. The proponents of Deep Ecology believe that the conditions for the diversity of the myriad forms of life including the cultural diversity of human life need to be sustained.
7. The ideological change is mainly that of appreciating life quality rather than adhering too increasingly higher standards of living. The existing pattern of labour and consumption are creating dissatisfaction among people. Quality of life leads to increase in happiness, vitality and joy.

8. Those who subscribe to the foregoing points have an obligation directly or indirectly to try to implement the necessary changes.

Implications of Deep Ecology

Deep Ecology offers a solution to the ongoing environmental crisis through a radical shift in human consciousness. A fundamental change is required in the way people relate to various facets of environment. Deep Ecology has shaped the environmental debate among leading activists and policy makers—from President Al Gore to Dave Foreman, cofounder of Earth First! Deep Ecology movement believes that protection of planet Earth from destruction requires a shift from an anthropocentric to a spiritual/ecocentric value orientation. It should be understood that ecological health of the planet and well being of the people is very much dependent on the intrinsic value of wild ecosystem and species. Deep Ecology movement considers it necessary that industrial activities must be significantly scaled down, consumption lifestyles must be changed, population pressure should be reduced, and remaining wildlife and wild ecosystem should be protected and restored by humanity.

Gaia Hypothesis

Throughout history all societies recorded their belief that Earth is a living organism. This belief was challenged at the beginning of the seventeenth century when the Cartesian/Newtonian scientific worldview that conceived the world in terms of a machine became popular. James Lovelock in 1970s revived the world-as-organism image and advocated a sophisticated cybernetic theory and referred it as the Gaia hypothesis (Lovelock, 1979).

A number of controversial philosophical and environmental implications were derived by Lovelock from the Gaia hypothesis:

1. Gaia has "vital organs" that need to be protected but

claimed that the Earth is very resilient and is capable of withstanding excessive abuse. For instance, Lovelock (1979) claimed that even the nuclear war is also not able to seriously damage Gaia because Gaia is open to unlimited development.

2. Lovelock claimed that the evolution of humans with high technological inventiveness and his increasingly subtle communication network has enhanced Gaia's range of perception. Gaia is through us awake and aware of herself. This hypothesis is consistent with Teilhard's (1959) metaphysics, which posits that the consciousness of the world is located exclusively in humans.
3. Gaia hypothesis and speculations treat Humans as the intelligence of Gaia. Perfecting Nature (Gaia) through genetic engineering, a takeover of the Earth's revolutionary processes, and of global development and management schemes based on vast electronic satellite communications systems (Sessions, 1995). A prolific New Age thinker Paul Vajk (1978) wrote "should we find it desirable, we will be able to turn the Sahara Desert into farms and forests, or remake the landscape of New England. We are the legitimate children of Gaia. We need not be ashamed that we are altering the landscapes and ecosystems of Earth."

The views adopted in the book entitled "Gala: An Atlas of Planetary Management" authored by Norman Meyers (1984) is endorsed by Lovelock in his Gaia hypothesis. The author calls for the preparation of an inventory of the Earth's "resources" as a basis for efficient global planning and exploitation. According to Meyers (1984) the rain forest may constitute a "gene library" for humans; the oceans are a "remarkable rich resource realm". A major solution to the environmental crisis will be provided by genetic engineering. Furthermore, Meyers believes that "with the power of life in our hand, we could make forests spring up on bare lands.... redeploys the genetic wealth of evolution to craft organisms to work in partnership with us.... It is time for humanity to use this power, and use it well." Finally, it seems that the ecological crisis move not towards the preservation of nature, but to the replacement of nature.

According to Morris Berman (1987) the development and interpretation forwarded by Meyers with regard to Gaia hypothesis is an example of "cybernetic dream." A noted New Age historian Thompson (1991) feels optimistic about the "cybernetic dream." He sees the possibility of "downloading" the human mind into a computer. Thompson believes that the United States is going to lead the way "from a literary and European culture to a planetary and electronic one." In other words the US is going to prepare the world for a new global culture in which nature will be replaced by a wholly artificial environment.

The New Age proponent David Spangler (1984) claims, "the earth was entering a new cycle of evolution, which would give birth to new civilization." Jerry Mandler (1995) visualizes the New Age as combining with Gaia hypothesis to provide a "techno-religious rationalization for ......Space-Colonization-as-Evolution" where we are (Gaia's) brain. Mander points out that "a lot of New Age people love the idea of leaving the planet, being chosen by evolution to be its personal astronauts escaping into space. The prospect of personally fulfilling nature's evolutionary design is so thrilling an idea that its advocates don't see, or don't care, that it is only a modern-day continuation of manifest Destiny, with the same outcome." Teilhard (1959) asserts that human progress through technology is inevitable and the end result will be seen in a form of "planetary culture" and or "global civilization". A number of other scholars such as Alvin Toffler (1990) believe that a "third wave" of global planetization involving a "technosphere", "infosphere", and so forth is likely. The New Age movement derives its support and inspiration from scholar and technologist Fuller (1957) who coined the phrase "spaceship Earth". According to Fuller (1957) the humans are now in a position to take control of the biological systems of the Earth. As a technological utopia, he asserts that there is no human population problem with modern technology there can be four billion billionaires on the planet! Engineers and computers, together with cybernetics and systems theory, will provide purely technological solution to the world's problems. A number of scholars have severely criticized the New Age thinkers and stated that they are busy promoting an abstract and disembodied perception of reality and that modern technological society is becoming addicted to a pure mental process which he labels as "computer consciousness" (Berman, 1886).

Summary

This chapter deals with various environmental issues that concern people. Various behavioural interventions to arrest the environmental destructive behaviour and promote pro-environmental tendencies are discussed. With growing awareness of disturbed man-environmental relationship various people centered initiatives and movements are carried out to protect and conserve the environment. Two case studies—Chipko and Narmada Andolan—treated as social movements are comprehensively discussed. Finally, two competing theoretical perspective dealing with environmental problems are discussed. The Deep ecology provides some prescriptions about the way one should behave with the environment to avert crisis. In contrast, the proponent of Gaia framework strongly feels that there is no crisis and human being is capable to dealing with environmental problems.

REFERENCES

Agarwal, A., Mant, D'.,& Samarth, V. (1987). *The Fight for Survival: People's Action for Environment*. New Delhi: Centre for Science and Environment.

Allen, G.H. (1972). How deep is environmental awareness? *Journal of Environmental Education*, 3, 1-3.

Asch, J., & Shore, B. (1975). Conversation behavior as the outcome to environmental Education. *Journal of Environmental Education*, 6, 25-33.

Baltes, M.M., & Hayward, S.C. (1976). Application and evaluation of strategies to reduce pollution: Behavioral control of littering in a football stadium. *Journal of Applied Psychology*, 61, 501-506.

Berman, M. (1986). The Cybernetic Dream of the Twenty-first Century. *Journal of Humanistic Psychology*, 26, 24-51.

Bhatt, C.P. (1987). The Chipko Andolan: Forest Conservation based on People's power. In Anil Agarwal, Durry D' Monte and Ujwala Somarth (Eds.), *The Fight for Survival*. New Delhi: Centre for Science and Environment.

Capra, F. (1995). Deep Ecology: A new paradigm. In George Sessions' (Ed.), *Deep Ecology of the 21st century* (pp. 19-25).

Carson, R. (1962). *The Silent Spring*. Boston: Houghton: Miffin.

Cass, R., & Edney, I.J. (1978). The Common Dilemmas: A simulation testing the effects of resource visibility and territorial division. *Human Ecology*, 6, 371-386.

Clark, R.N., Hendee, J.C., & Burgess, R.L. (1972). The experimental control of littering. *Journal of Environmental Education*, 4, 22-28.

Cohen, M.R. (1973). Environmental information vs. environmental attitudes. *Journal of Environmental Education*, 5, 5-8.

Cone, J.D., & Hayes, S.C. (1978). Applied behavioral analysis and the solutions of environmental problems. In J.F. Wohlwill & I. Altman (Eds.), *Human Behavior and the Environment: Advances in Theory and Research* (Vol. 2), New York: Plenum.

Cone, J.D., & Hayes, S.C. (1980). *Environmental Problems/Behavioral Solutions*. Monterey: Brooks/Cole.

Dharchaudhuri, I. (2003). A river dam(n)ed. *Hindustan Times*, 13th August, p. 2.

Everett, P.B. (1977). *A behavioral Science Approach to Transportation Systems Management*. Unpublished manuscript.

Fox, S. (1991). Self and world: A transpersonal, ecological approach. *Revision*, 1, 13.

Foxx, R.M., & Hake, D.F. (1977). Gasoline conservation: a procedure for managing and reducing the driving of college students. *Journal of Applied Behavioral Analysis*, 10, 61-74.

Fox, W. (1986). Approaching Deep Ecology: A response to Richard Sylvan's Critique of Deep Ecology. *Environmental Studies Occasional Paper* 20, Hobart: University of Tasmania.

Fuller, B. (1957). Man with the chronofile. *Sunday Review*, April 14-18.

Geller, E.S., Winnett, R.A., & Evert, P.B. (1982). *Preserving the Environment: New Strategies for Behavioral Change*. New York: Pergamon.

Hake, D.F., & Foxx, R.M. (1978). Promoting gasoline conservation: The effects of reinforcement schedules, a leader and self-recording. *Behavior Modification*, 2, 339-369.

Hardin, G. (1968). The tragedy of the commons. *Science*, 162, 1243-1248.

Hayes, S.C., & Cone, J.D. (1977). Reducing residential electrical energy use: Payments, Information and feedback. *Journal of Applied Behavioral Analysis*, 10, 425-435.

Ingram, R.E., & Geller, E.S. (1975). A community integrated behavior modification approach to facilitating paper recycling. *JSAS Catalog of Selected Documents in Psychology*. 5, 327.

Left, H.I. (1974). *Experience, Environment and Human Potentials*. New York: Oxford University Press.

Lingwood, D.A. (1971). Environmental education through information seeking: The case of an environmental teach in. *Environment and Behavior*, 3, 220-262.

Lovelock, J. (1979). *Gaia: A New Look at Life on Earth*. New York: Oxford University Press.

Mandler, J. (1995). Leaving the Earth Space Colonies, Disney and Epcot. In G. Sessions (Eds.) *Deep Ecology for the 21st century*, Massachusetts: Shambhala Publications.

McLaughlin, A. (1995). The heart of deep ecology. In George Sessions (Ed.), *Deep Ecology for the 21st Century* (pp. 85-94).

Meyers, N. (1984). *Gaia: An Atlas of Planet Management*. Garden City, N.Y.: Anchor Press.

Meyers, A.E., Artz, L.M., Craighead, W.E. (1976). The effects of instructions, incentives, and feedback on a community problem: Dormitory noise. *Journal of Applied Behavioral Analysis*, 9, 445-457.

Mishra, A.B. (1987). Mining a hill and undermining a society: The case of Gandhamardan. In A. Agarwal, D.' Minte & U. Samarth (Eds.), *The Fight for survival: Peoples' Action for Environment*, pp. 125-144.

Naess, A. (1995). The shallow and the deep, long-range ecological movements. In G. Sessions (Ed.) *Deep Ecology for the 21st Century*, pp. 151-155.

Olsen, M.E. (1981). Consumers' attitudes toward energy conservation. *Journal of Social Issue*, 37, 108-131.

Rankin, R.E. (1969). Air pollution, control and public apathy. *Journal of the Air Pollution Control Association*, 19, 565-569.

Reichel, D.A., & Geller, E.S. (1980). *Group versus Individual Contingencies to Conserve Transportation Energy*. Paper presented at the Southwestern Psychological Association meeting, Washington, D.C.

Sadik, N. (1992). *The State of World's Population*. New York: United Nations Population Fund.

Schmidt, C.W., & Ulrich, R.E. (1969). Effects of group contingent events upon classroom noise. *Journal of Applied Behavioral Analysis*, 2, 171-179.

Seligman, C., Kriss, M., Darley, J.M., Fazio, R.H., Becker, L.J., & Pryor, J.B. (1979). Predicting summer energy consumption from home owners' attitudes. *Journal of Applied Social Psychology*, 9, 70-90.

Sessions, G. (1995). *Deep Ecology for the Twenty-First Century*. Boston: Shanbhala.

Socolow, R.H. (1978). *Saving Energy in the Home*. Cambridge, Mass: Ballinger Publishing Company.

Spangler, D. (1984). *Emergence: The Rebirth of the Sacred*. New York: Dell.

Teilhard, D.C. (1959). *The Phenomenon of Man*. New York: Harper and Row.

Thompson, W.I. (1991). *The American Replacement of Nature*. New York: Doubleday.

Toffler, A. (1990). *Powershift*. New York: Bantam Books.

Vajk, P. (1978). *Doomsday Has Been Cancelled*. Menlo Park, California: Peace Publishers.

White, L. (1967). The historical roots of our ecologic crisis. *Science*, 155, March 10.

Wilson, E.O. (1992). *The Diversity of Life*. Cambridge: Harward University Press.

Wilson, B.J., & Hopkins, B.L. (1973). The effects of contingent music on the intensity of noise in junior high school economics classes. *Journal of Applied Behavior Analysis*, 6, 269-275.

Winston, B.J. (1974). The relationship of awareness to concern for environmental quality among selected high school students. *Dissertation Abstracts International*, 35A, 3412.

APPENDIX

INDWELLING CROWDING EXPERIENCE SCALE

Instructions

This questionnaire is related to some aspects of your home environment. We understand each family has its inbuilt provisions and traditions. The items in this questionnaire are designed to assess how the young family members like you perceive the climate of your home environment in terms of involvement, emotional support, possibility of intellectual growth, independence, norms and values, social orientation and privacy. We are interested to know how you and your family members influence each other. There is no model or standard answers to these items. You are required to scale your reactions on the 5-point Likert type rating scale given below.

Involvement

1. To what extent you willingly engage yourself in the daily household activities?
   Not at all   A little   Somewhat
   Quite a bit   Very much
2. How much of your spare time is spent with the members of your family?
   Very often   Often   Sometimes
   Rarely   Never
3. I help my family to clear out the accumulated household work.
   Not at all   A little   Somewhat
   Quite a bit   Very much

4. How much aware you are about the goings on in your family?
   Not at all A little Somewhat
   Quite a bit Very much

Emotional Support

5. I share my problems with the members of my family.
   Not at all A little Somewhat
   Quite a bit Very much
6. How free you feel to communicate with the members of your family?
   Very often Often Sometimes
   Rarely Never
7. How often members of your family attend to your difficulties?
   Very often Often Sometimes
   Rarely Never
8. How often you are concerned about the welfare of others in your family?
   Very often Often Sometimes
   Rarely Never

Independence

9. Members of my family don't impose their decisions on me.
   Not at all A little Somewhat
   Quite a bit Very much
10. I feel free to decide things for myself.
   Not at all A little Somewhat
   Quite a bit Very much
11. I feel elders impose their feelings on me.
   Not at all A little Somewhat
   Quite a bit Very much
12. I am free to go whenever and wherever I wish.
   Not at all A little Somewhat
   Quite a bit Very much

13. To what extent do your family members inhibit your friend's frequent visit at your home?
    Very often Often Sometimes
    Rarely Never
14. To what extent you feel that you are at your liberty to study or not?
    Not at all A little Somewhat
    Quite a bit Very much

Traditional Social Orientation

15. How often family members adhere to the traditions?
    Very often Often Sometimes
    Rarely Never
16. I anticipate some problems when I ask permission to go for a movie.
    Not at all A little Somewhat
    Quite a bit Very much
17. I feel like breaking the rigid traditional norms of my family.
    Very Often Often Sometimes
    Rarely Never

Intellectual Growth

18. To what extent the elders of your family encourage your independent thinking?
    Very often Often Sometimes
    Rarely Never
19. To what extent your family members are concerned about your studies?
    Not at all A little Somewhat
    Quite a bit Very much
20. I am encouraged to read magazines other than prescribed books by my elders.
    Very often Often Sometimes
    Rarely Never
21. My elders check me when I ignore studies.
    Very often Often Sometimes
    Rarely Never

22. Elders of my family never complain why I am not studying.
    Not at all A little Somewhat
    Quite a bit Very much
23. My family members encourage me towards extracurricular activities.
    Very often Often Sometimes
    Rarely Never

Order and Organization

24. I am subjected to harsh words when I am late returning home.
    Very often Often Sometimes
    Rarely Never
25. Elders of my family never inquire about my personal activities.
    Not at all A little Somewhat
    Quite a bit Very much

Student's Influence

26. To what extent parents take advice from you?
    Very often Often Sometimes
    Rarely Never
27. To what extent you are a party in all decision making?
    Very often Often Sometimes
    Rarely Never
28. My family members respect my feelings.
    Not at all A little Somewhat
    Quite a bit Very much
29. I am able to convince my elders about my idea, which is contrary to their own ideas?
    Not at all A little Somewhat
    Quite a bit Very much
30. I influence my parents by my advice.
    Not at all A little Somewhat
    Quite a bit Very much

31. My advice has a weightage in my family.
    Very often Often Sometimes
    Rarely Never
32. My advice is welcomed in interpersonal matters.
    Not at all A little Somewhat
    Quite a bit Very much
33. My advice is welcomed in economic matters.
    Not at all A little Somewhat
    Quite a bit Very much

Privacy

34. I enjoy as much privacy as I want at my home.
    Not at all A little Somewhat
    Quite a bit Very much
35. I am able to read desired literature at home when I want to.
    Very often Often Sometimes
    Rarely Never
36. To what extent you feel that there is no room where you can enjoy privacy?
    Very Often Often Sometimes
    Rarely Never

Source: Nagar, D. (1998). ***Human reactions to crowding***. Jaipur: Printwell Publishers & Distributors.

Residential Crowding Experience Scale

1. How spacious is your home/apartment?
2. Do you feel that each member in your home has sufficient amount of space?
3. How satisfied are you with the space available in your home?
4. To what extent does the amount of space interfere with your personal activities?
5. How uncomfortable are you with the space available in your home?

6. To what extent do you have a place to call your own in your home?
7. How often do others in your home make it difficult for you to carry out your daily routine?
8. How often do you spend your time outside your home because of insufficient space?
9. How often are you able to have as much privacy as you desire in your home?
10. To what extent are the residents of your home the kind of people with whom you can communicate easily?
11. To what extent are the residents of your home the kind of people with whom you can share your personal problems?
12. How friendly are the other residents of your home?
13. How often do you spend time with the other residents of your home?
14. To what extent do you trust other residents of your home?
15. How often do other residents of your home provide emotional support when you need it?
16. How often do others in your home make unwanted demands on you?
17. How often do you feel ignored by other residents of your home?
18. How often are you criticized by other residents of your home?
19. How hostile are the other residents of your home?
20. How often do residents enter your room without knocking?
21. How difficult is it to get along with other residents of your home?
22. How often are you disturbed by the playing of radios, TVs, or other such activities?
23. How often are you disturbed by outside noise?
24. How often are you disturbed by noise from people and their activities inside the home?
25. How often do residents fail to switch off the lights before going to bed?
26. How often do residents of your home ignore your request for quiet?

Note: All Items are to be rated on a 5 point Likert Type Rating Scale.

Factor 1: Space Satisfaction: Items 1–9.
Factor 2: Positive or supportive relationship: Items 10–15.
Factor 3: Negative or disruptive relationship: Items: 16–21.
Factor 4: Uncontrolled disturbance: Items 22–26.

Source: Nagar, D., & Paulus, P.B. (1997). *Journal of Community Psychology*, 7, 303-319 (John Wiley and Sons Ltd.).

Feeling of Crowding Scale

The scale contains 18 items tapping the feeling in high density conditions. Each item has a 5 point scale with end points 'completely' (1) to 'not at all' (5).

1. Do you feel crowded in your home?
2. Do you feel that you have a small space to live?
3. Do you have enough space for your children to study?
4. Do you feel disturbed due to constant noise in your house?
5. Do your place is disturbed due to noises of other houses or of street?
6. Do you like noise of the market?
7. Do you feel perturbed in the crowd?
8. Do you feel irritated because of crowd around you?
9. Do you feel disturbed because of guest because you have small space?
10. Do you feel nauseating due to dirty surrounding of your house?
11. Do you feel nauseating due to uncleanness because of crowd around your house?
12. Do you feel that because of small space your house does not remain clean?
13. Do you often feel that in case you get bigger space to live you will leave this house?
14. Do you feel that if you have enough money you will buy a bigger house in less crowded place?
15. Do you feel that in this ward living is becoming difficult day-by-day because of increased population?
16. Do you feel that you and your wife get full opportunity to enjoy privacy in this place?

17. Do you think people from neighbouring house disturb your privacy?
18. Do you feel disturbance from people walking on the street near your house?

Lack of Space: Items 2, 3, 9, 12, 13, 14
External Disturbance: items 1, 10, 11, 15, 17, 18
Mood Disturbance: Items 7, 8
Noise Disturbance: Items 4, 5, 6

Objective Crowding Measures:
A 6-item measure was developed to assess the extent to which a person is subjected to real crowding in day-to-day life. Each item was rated on 5-point scale ranging from 'Never' (1) to 'Always' (5).

1. I have to work in the crowded condition.
2. Most of my time is spend in the congested area of the town.
3. Crowd is there in my house.
4. I have to pass through congested areas.
5. There is lot of noise around my house.
6. There is lot of noisy traffics near my house.
   Populousness: Items 1-4.
   Noise: Items 5, 6.

Competition Tolerance

In our day-to-day life we face different situations like appearing in job related interviews, standing in queue to get scarce items, good crop etc. In all these situations we do some work like standing in queue ahead of time, yourself relaxing and sending someone else to get needed items, getting the items from open market instead of standing in queue. In all these situations we have to do one thing at a given point of time.

Given below are descriptions of 18 day-to-day competitive situations. You are required to keep yourself in each situation and select one of the four alternatives which best represent your position. The answers are arranged in a sequence to represent the most to least competitive reaction out of which respondents can opt for anyone. In addition, the participants are allowed to mention any other alternative they think true for them.

1. Suppose you are contesting election and your opponent is likely to win then....
   a. You will intensify your campaign work.
   b. Campaign work will go on at the same pace.
   c. Make an agreement with the opponent.
   d. Withdraw your name.
2. Suppose you are planning to start a business then you come to know that someone else has just started the same business what you were planning to start than—
   a. The same business will be carried out on a larger scale.
   b. The business will be carried out as per the previous plan.
   c. To plan business which others are not doing in your city.
   d. To drop the idea of doing business of any type.
3. Suppose you play a match and your team is defeated then what reasons you will give to others for the defeat....
   a. In fact the other team played very well.
   b. Our team although play well but today was not in form.
   c. All players of our team played were disappointed in the game.
   d. Players of the opponent team played against the rule and won the game.
4. If in your field or work area various competitive programmes are being organized where participation is entirely voluntary, than,....
   a. You will always participate.
   b. You will sometimes participate.
   c. You will not always participate.
   d. Your will never participate.
5. Suppose you very much need a scooter and there is a booking in your city and you expect a long queue than...
   a. You will yourself stand in the queue and do the booking.

b. Will use other method and try for a booking.
c. Purchase old scooter.
d. Drop the idea of purchasing a scooter.

6. If there are only two candidates for the job interview and you have come to know that the other candidate is more competent than you, then....
   a. Appear in the interview with more preparation.
   b. Appear in the interview but will not give the interview well.
   c. Appear in other interview where the candidates are not very able.
   d. Not appear in the interview.
7. If you have to participate in a competition, then you....
   a. Will easily adjust.
   b. Will adjust well in only those situations when you know that your work will be much better than others.
   c. Will feel anxious, tense and uncomfortable.
   d. Will participate in competition in only those situation where participation is compulsory.
8. In your office there are a few persons much more able than you. The work also requires skill and effort than you...
   a. Will work more hard and will try to develop more skill.
   b. Will not develop any skill but will develop ways to make the officers happy.
   c. Will work as usual.
   d. Will leave the work and try to search for job that requires less ability.
9. If your neighbour has many luxury things and his standard of living is higher than you, than you will—
   a. Will work hard and try to increase your income and maintain the same standard as that of others.
   b. Will try to enhance your status going beyond your income.
   c. Will not really enhance the status but will try to show off the high standard.
   d. Will not care to enhance the status and will continue to maintain the status quo.
10. You enjoy congenial relationship with your colleagues in your job. Suppose you get job at other place with increased

salary but there your performance will always be compared with others, then....

a. Join the new job.
b. Will try to get increased salary in the same job.
c. Will do the same job happily without increase in the salary.
d. Will do the same job amidst disappointment without increase in salary.

Prestige: Items: 1, 3, 4, 5, 6, 7, 8
Economic: Items 2, 9, 10.

Note: Eight situations were dropped because factor loadings were low and insignificant.

Source: Jain, U. (1997). *The feeling of crowding and its psychological consequences.* New Delhi: Har-Anand Publications Pvt. Ltd.

BIBLIOGRAPHY

Agarwal, A., & Narain, S (eds.). (1984-85). *The State of India's Environment: A Citizens' Report* (Second), New Delhi: CSE.

Agarwal, A., & Saxena, A.K. (2003). *Psychological Perspectives in Environment and Developmental Issues*. New Delhi: Concept.

Agarwal, A., Mant, D'.,& Samarth, V. (1987). *The Fight for Survival: People's Action for Environment*. New Delhi: Centre for Science and Environment.

Aiello, J., Nicosia, G., & Thompson, D. (1979). Physiological, social and behavioural consequences of crowding on children and adolescents. *Child Development*, 50, 195-202.

Aiello, J.R. (1987). Human spatial behaviour. In D. Stokols and I. Altman (Eds.), *Handbook of Environmental Psychology*, John Wiley & Sons, New York.

Aiello, J.R., & Aiello, T.D. (1974). The development of personal space: Proxemic behaviour of children 6 through 16. *Human Ecology*, 2, 177-189.

Aiello, J.R., & Cooper, R.E. (1979). *Personal Space and Social Affect: A Developmental Study*. Paper presented at the Meeting of the Society for Research in Child Development, San Francisco.

Aiello, J.R., & Thompson, D.E. (1980). When compensation fails: Mediating effects of sex and locus of control at extended interaction distances. *Basic and Applied Social Psychology*, 1, 65-82.

Aiello, J.R., Baum, A., & Gormley, F. (1981). Social determinants of residential crowding stress. *Personality and Social Psychology Bulletin*, 7, 643-649.

Aiello, J.R., Epstein, Y., & Karlin, R. (1975). Effects of crowding on electrodermal activity. *Sociological Symposium*, 14, 42-57.

Alland, A. (1972). *The Human Imperative*. New York: Columbia University.

Allen, G.H. (1972). How deep is environmental awareness? *Journal of Environmental Education*, 3, 1-3.

Altman, I. (1975) *Environment and Social Behavior: Privacy, Personal Space, and Territory and Crowding*. Monterey, California: Brooks Cole.

Altman, I. (1976). Environmental psychology and social psychology. *Bulletin of Social and Personality Psychology*, 2, 96-113.

Altman, I. (1978). Crowding: Historical and contemporary trends in crowding research. In A. Baum & Y.M. Epstein (Eds.), *Human Response to Crowding*. Hillsdale, N.J.: Erlbaum.

Altman, I., & Chemers, M. (1980). *Culture and Environment*. Monterey, CA: Brooks/Cole.

Altman, I., & Rogoff, B. (1987). World's view in psychology: Trait, interactional, organismic, transactional perspectives. In D. Stokols, & I. Altman (Eds.), *Handbook of Environmental Psychology* (Vol. 1. pp.7-40). New York: John Wiley.

Altman, I., & Stokols, D. (1987). *Handbook of Environmental Psychology*. New York: Wiley.

Altman, I., Taylor, D.A., & Wheeler, L. (1971). Ecological aspects of group behaviour in social isolation. *Journal of Applied Social Psychology*, 1, 76-100.

Appleyard, D. (1970). Styles and methods of structuring a city. *Environment and Behavior*, 2, 100-118.

Appleyard, D. (1976). *Planning a Pluralist City*. Cambridge, MA: MIT Press.

Appleyard, D. (1979). The Environment as a Social Symbol. *Journal of the American Association*, 45, 143-153.

Ardey, P. (1966). *The Territorial Imperative*. New York: Atheneum.

Ardey, R. (1970). *The Social Contract*. New York: Atheneum.

Arenson, D.D. (1977). Reactions to invasions of market seats at a racetrack. *Social Behavior and Personality*, 5, 225-228.

Argyle, M., & Cook, M. (1976). *Gaze and Mutual Gaze*. Cambridge, England: Cambridge University Press.

Argyle, M., & Dean, J. (1965). Eye contact, distance and affiliation. *Sociometry*, 28, 289-304.

Asch, J., & Shore, B. (1975). Conversation behaviour as the

outcome to environmental Education. *Journal of Environmental Education*, 6, 25-33.

Asthana, H., & Jain, U. (1986). Alienation in Bhopal MIC gas victims. *Paper Presented At XI World's Congress of Sociology*, New Delhi: August.

Averill, J. (1973). Personal control over aversive stimuli and its relationship to stress. *Psychological Bulletin*, 80, 286-303.

Baird, J.C. (1979). Studies of the cognitive representation of spatial relations: An overview. *Journal of Experimental Psychology General*, 108, 90-91.

Baird, J.C., Wagner, M. & Noma, E. (1981). Impossible cognitive spaces. *Geographical Analysis*, 14, 204-216.

Baird, L.L. (1969). Big school, small school: A critical examination of the hypothesis. *Journal of Educational Psychology*. 60, 253-260.

Baltes, M.M., & Hayward, S.C. (1976). Application and evaluation of strategies to reduce pollution: Behavioural control of littering in a football stadium. *Journal of Applied Psychology*, 61, 501-506.

Barefoot, J.C., Hoople, H., & McClay, D. (1972). Avoidance of an act which would violate personal space. *Psychonomic Science*, 28, 205-206.

Barker, R. (1960). *Ecology and Motivation*. Nebraska Symposium on Motivation.

Barker, R. (1976). Planning for environmental indices: Observer appraisal of air quality. In K. Craik & E. Zube (eds.), *Perceiving Environmental Quality* (pp. 175-204). New York: Plenum.

Barker, R., & Gump, P. (1964). *Big School, Small School*. Stanford University Press.

Barker, R.G. & Wright, H. (1955). *Midwest and its Children*. New York: Row and Peterson.

Barker, R.G. (1960). Ecology and motivation. In M.R. Jones (Ed.), *Nebraska Symposium on Motivation* (Vol. 8). Lincoln: University of Nebraska Press.

Barker, R.G. (1968). *Ecological Psychology: Concepts and Methods for Studying the Environment and Human Behavior*. Stanford, CA: Stanford University Press

Barker, R.G. (1979). Setting of a professional lifetime. *Journal of Personality and Social Psychology*, 37, 2137-2157.

Barker, R.G. (Ed.). (1978). *Habitats, Environments, and Human Behavior: Studies in Ecological Psychology and Eco-Behavioral Science from the Midwest Psychological Field Station*, 1947-1972. San Francisco: Jossey-Bass.

Barker, R.G., & Gump, P.V. (1964). *Big School, Small School: High School Size and Student Behaviour.* Stanford, CA: Stanford University Press.

Barker, R.G., & Wright, H.F. (1949). Psychological ecology and the problem of psychosocial development. *Child Development*, 20, 131-143.

Baron, R.A., & Bell, P.A. (1976). Physical distance and helping: Some unexpected benefits of "crowding in" others. *Journal of Applied Social Psychology*, 6, 95-104.

Baron, R.M., Mandel, D.R., Adams, C.A., & Griffitt, L.M. (1976). Effects of social density in university residential environments. *Journal of Personality and Social Psychology*, 34, 434-446.

Batra, A., Muhar, I., & Bhatia, P. (1990). Environmental noise and mental work. *Indian Journal of Psychology*, 65, 45-52.

Baum, A., & Gatchel, R. (1980). *Cognitive Determinants of Reaction to Uncontrolled Events: Development of Reactance and Learned Helplessness.* Unpublished manuscript, Uniformed Services, University of Health Science.

Baum, A., & Greenberg, C.I. (1975). Waiting for a crowd: the behavioural and perceptual effects of anticipated crowding. *Journal of Personality and Social Psychology*, 32, 671-679.

Baum, A., & Koman, S. (1976). Differential response to anticipated crowding: Psychological effects on social and spatial density. *Journal of Personality and Social Psychology*, 34, 526-536.

Baum, A., & Paulus, P. (1987). Crowding. In D. Stokols & I. Altman (Eds.), *Handbook of Environmental Psychology* (Vol. 1, pp. 553-570), New York, John Wiley.

Baum, A., & Valins, S. (1977). *Architecture and Social Behavior: Psychological Studies in Social Density*. Hillsdale, N.J.: Erlbaum.

Baum A., & Valins, (1979). Architectural mediation of residential crowding and control: crowding and the regulation of social contact. In L. Berkowitz (Ed.), *Advances in Experimental Psychology* (Vol. 12), pp. 131-175), New York: Academic Press.

Baum, A., Aiello, J., & Calesnick, L. (1978). Crowding and personal control: Social density and the development of learned helplessness. *Journal of Personality and Social Psychology*, 36, 1000-1011.

Baum, A., Fleming, R., & Davidson, L. (1983). Natural disaster and technological catastrophe. *Environment and Behaviour*, 15, 333-354.

Baum, A., Greenberg, C.I. (1975). Waiting for a crowd: The behavioural and perceptual effects of anticipated crowding. *Journal of Personality and Social Psychology*, 32, 671-679.

Baum, A., Grunberg, N.E., & Singer, J.E. (1982).The use of psychological and neuroendocrinological measurement in the study of stress. *Health Psychology*, 1, 217-236.

Baum, A., Singer, J.E., & Baum, C. (1982). Stress and the environment. In G.W. Evans (Ed), *Environmental Stress* (pp. 15-44), New York: Cambridge University Press.

Baum, A., Singer, J.E., & Baum, C.S. (1981). Stress and the environment. *Journal of Social Issues*, 37, 4-35.

Baum, P., & Paulus, P.B. (1987). Crowding. In D. Stokols & I. Altman (Eds.), *Handbook of Environmental Psychology* (pp. 533-570). New York: Wiley.

Bechtel, R.B. (1977). *Enclosing Behavior.* Stroudburg, PA: Dowden, Hutchinson & Ross.

Bell, P.A., & Barnard, S.W. (1977). *Sex Differences in the Effects of Heat and Noise Stress on Personal Space Permeability.* Paper presented at the meeting of the Rocky Mountain Psychological Association, Albuquerque, May.

Bergman, B.A. (1971). *The Effects of Group Size, Personal Space, and Success-Failure on Physiological Arousal, Test Performance, and Questionnaire Response.* Unpublished Doctoral dissertation, Temple University, Philadelphia.

Berman, M. (1986). The Cybernetic Dream of the Twenty-first Century. *Journal of Humanistic Psychology*, 26, 24-51.

Beutell, A.W. (1934). An analytic basis for a lighting code. *Illuminating Engineering*, 27, 5-16.

Bhatia, P., & Muhar, I. (1988). Noise sensitivity and mental efficiency. *Psychologia*, 31, 163-169.

Bhatt, C.P. (1987). The Chipko Andolan: Forest Conservation based on People's power. In Anil Agarwal, Durry D' Monte and Ujwala Somarth (Eds.), *The Fight for Survival*. New Delhi: Centre for Science and Environment.

Bhattacharya, S.K., Rao, A., & Chatterjee, S. (1978). Comparison of motor-manual performance in quiet and noisy environments. *Indian Journal of Medical Research*, 68, 306-311.

Binet, M.A. (1894). Reverse illusions of orientation. *Psychological Review*, 1, 337-350.

Blaut, J.M. & Stea, D. (1971). Studies of geographic learning. *Annals of the Association of American Geographers*, 61, 387-393.

Bonnes, M., & Secchiaroli, G. (1995). *Environmental Psychology: A Psycho-Social Introduction*. London: Sage.

Booraem, C.D., & Flowers, J.V. (1972). Reduction of anxiety and personal space as a function of assertion training in severely disturbed neuropsychiatric inpatients. *Psychological Reports*, 30, 923-929.

Booth, A., & Johnson, D. (1975). *The Effects of Crowding on Child Health and Development*, Mimeo.

Booth, A., & Welch, S. (1973). *The Effects of Crowding: A Cross-National Study*. Unpublished manuscript, Ministry of State for Urban Affairs, Ottawa, Canada.

Brehm, J. (1966). *A Theory of Psychological Reactance*. New York: Academic Press.

Briggs, R. (1972). *Cognitive Distance in Urban Space*. Unpublished doctoral dissertation, Ohio State University, Columbus.

Broadbent, D. (1957). Effects of noise on behaviour. In C.M. Harris (ed.), *Handbook of Noise Control*, New York: Academic Press.

Broadbent, D.E. (1971). *Decision and Stress*. New York: Academic Press.

Brolin, B. (1972). Experts planned Chandigarh but something has gone wrong. *Smithsonian*, 3, 56-62.

Bronzaft, A.L., & McCarthy, D. (1975). The effects of elevated train noise on reading Ability. *Environment and Behavior*, 7, 517-527.

Brown, B.B., & Altman, I. (1983). Territoriality, street form, and residential burglary: An environmental analysis. *Journal of Environmental Psychology*, 3, 203-220.

Brown, W. (1932). Spatial integrations in a human maze. *University of California Publications in Psychology*, 5, 123-134.

Brunswik, E. (1944). Distal focusing on perception. *Psychological Monographs*, 56, 1-48.

Brunswick, E. (1947). *Systematic and Representative Design of Psychological Experiments* .LA: University of California Press.

Bullinger, M. (1989). Psychological effects of air pollution on healthy residents: A Time Series approach. *Journal of Environmental Psychology*, 9, 103-118.

Burgess, J.W. (1983). Interpersonal spacing behaviour between surrounding nearest neighbors reflects both familiarity and environmental density, *Ethology and Sociobiology*, 4, 11-17.

Burgess, J.W., & McMurphy, D. (1982). The development of proxemics spacing behaviour: Children's distances to surrounding playmates and adults change between 6 months and 15 years of age. *Developmental Psychobiology*, 15, 557-567.

Burnette, C., Moore, G.T., & Simek, (1973). A role-oriented approach to problem Solving groups. In W. F. E. Preiser (Ed.), *Environmental Design Research* (Vol. 1). Stroudsburg, Pa: Dowden, Hutchinson & Ross.

Burns, TY. (1964). Non-verbal communication. *Discovery*, 31-35.

Burroughs (Eds.), *Introduction to Applied Psychology*. London: Scott Forsman and Company.

Burroughs, W.I. (1989). *Applied environmental psychology*. In W.L. Gregory & W.J.

Calhoun, J. (1962). Population density and social pathology. *Scientific American*, 206, 139-148.

Calhoun, J. (1973). Death squared: The explosive growth and demise of a mouse population. *Proceedings of the Royal Society of Medicine*, 66, 80-88.

Calhoun, J.B. (1962). Population density and social pathology. *Scientific American*, 206, 157-163.

Calsyn, R. (1976). Group responses to territorial intrusion. *Journal of Social Psychology*, 40, 51-58.

Campbell, D.T., & Stanley, J.C. (1963). *Experimental and Quasi-Experimental Designs for Research.* Chicago: Rand McNally.

Campbell, J. (1983). Ambient stressors. *Environment and Behaviour*, 15, 355-380.

Canter, D. (1968). *The Measurement and Meaning and Architecture.* Unpublished Manuscript. Glascow Building Performance Research Unit.

Canter, D. (1969). A intergroup comparison of connotative dimensions. *Environment and Behaviour*, 1, 37-48.

Canter, D. (1983). The purposive evaluation of places: A facet approach. *Environment and Behavior*, 15, 659-698.

Cantor, D. (1970). *Architectural Psychology*. London: Royal Institute of British Architects.

Caplow, T., & Forman, R. (1950). Neighborhood interaction in a homogeneous community. *American Sociological Review*, 15, 357-366.

Cappella, J.N., & Greene, J.O. (1982). A discrepancy arousal explanation of mutual Influence in expressive behaviour in adults and infant-adult interaction. *Communication Monographs*, 49, 89-114.

Capra, F. (1995). Deep Ecology: A new paradigm. In George Sessions' (Ed.), *Deep Ecology of the 21st century* (pp. 19-25).

Carson, R. (1962). *The Silent Spring*. Boston: Houghton Miffin.

Carstairs, G. (1969). Overcrowding and human aggression. In H.D. Graham & T.D. Curr (Eds.), *Violence in America: Staff Report to the National Commission on the Causes and Prevention of Violence.* New York: Signet Books.

Carstens, D.Y. (1985). *Site Planning and Design for the Elderly: Issues, Guidelines and Alternatives.* New York: Van Nostrand Reinhold.

Caspi, A., Bolger, N., & Eckenrode, J. (1987). Linking person and context in the daily Stress process. *Journal of Personality and Social Psychology*, 52, 184-195.

Caspi, A., Bolger, N., & Eckenrode, J. (1987). Linking persona and the context in the daily stress process. *Journal of Personality and Social Psychology*, 52, 184-195.

Cass, R., & Edney, I.J. (1978). The Common Dilemmas: A simulation testing the effects of resource visibility and territorial division. *Human Ecology*, 6, 371-386.

Cassell, J. (1971). Human consequences of population density. *Rapid Population Growth : Consequences and Policy Implications*, Baltimore: John Hopkins Press.

Cassidy, J. (1997). *Environmental Psychology: Behavior and Experience in Context.* Sussex: Psychology Press.

Cassidy, T. (1997). *Environmental psychology: Behavior and Experience in context. Psychology.* Press Publisher, UK.

Chattopadhyay, P.K., Som, B., Biswas, D. (1993). Air pollution and health hazards: An exploratory study. *Indian Journal of Clinical Psychology*, 20, 25-30.

Christensen, D., Carp, F., Cranz., & Wiley, J. (1982). Objective housing indicators as predictors of subjective evaluations of elderly residents. *Journal of Environmental Psychology*, 12, 225-236.

Christian, J. (1963). Endocrine adaptive mechanism and the physiologic regulation of population growth. In W. Mayor & R. Gelder (Eds.), *Physiological Mammalogy*,(Vol. 1) New York: Academic Press.

Christian, J., Flyger, V., & Davis, D. (1960). Factors in the mass mortality of a herd of Silka deer cervus nippen. *Chespapeake Science*, 1, 19-95.

Chuaqui, C., Lemkau, P., Legarrefa, A., Contresas, M. (1966). *Suicide in Santiago, Chile.* Public health reports, 81, 1100-1117.

Clark, R.N., Hendee, J.C., & Burgess, R.L. (1972). The experimental control of littering. *Journal of Environmental Education*, 4, 22-28.

Coffin, D., & Stokinger, H. (1977). Biological effects of air pollutants. In A.C. Stern (ed.), *Air Pollution* (3rd ed. Vol. 3, pp. 231-360), New York: Academic.

Cohen, M.R. (1973). Environmental information vs. environmental attitudes. *Journal of Environmental Education*, 5, 5-8.

Cohen, S. (1978). Environmental load and the allocation of attention. In A. Baum., J.E. Singer., & S. Valins (Eds.), *Advances in Environmental Psychology* (Vol. 1), N.J.: Erlbaum.

Cohen, S. (1980). Aftereffects of stress on human performance and social behaviour: A review of research and theory. *Psychological Bulletin*, 87, 578-604.

Cohen, S., & Syme, S. L. (Eds.). (1985). *Social Support and Health*. New York: Academic Press.

Cohen, S., & Weinstein, N. (1981). Non-auditory effects of noise on behaviour and Health. *Journal of Social Issues*, 37, 36-70.

Cohen, S., Evans, G., Krantz, D.S., Stokols, D.(1980). Physiological, motivational, and cognitive effects of aircraft noise on children: Moving from the laboratory to the field. *American Psychologists*, 35, 231-243.

Cohen, S., Glass, D.C., & Singer, J.E. (1973). Apartment noise, auditory discrimination and reading ability in children. *Journal of Experimental Social Psychology*, 9, 407-422.

Cohen, S.A. (1978). Environmental load the allocation of attention. In A. Baum, J.E. Singer, & S. Valins (Eds.), *Advances in Environmental Psychology* (Vol. 1). Hillsdale: Erlbaum.

Commoner, B. (1963). *Science and Survival*. New York; Viking.

Cone, J.D., & Hayes, S.C. (1978). Applied behavioural analysis and the solutions of environmental problems. In J.F. Wohlwill & I. Altman (Eds.), *Human Behavior and the Environment: Advances in Theory and Research* (Vol. 2), New York: Plenum.

Cone, J.D., & Hayes, S.C. (1980). *Environmental Problems/ Behavioral Solutions*. Monterey: Brooks/Cole.

Cook, M. (1970). Experimentation in orientation and proxemics. *Human Relations*, 23, 61-76.

Cook, T.D., & Campbell, D.T. (1976). The design and conduct of quasi-experiments and true experiment in field settings. In M.D. Dunette (Ed.), *Handbook of Industrial and Organizational Psychology* (pp. 223-316). Chicago: Rand McNally.

Cook, T.D., & Campbell, D.T. (1979). *Quasi-Experimentation: Design and Analysis Issues for Field Settings*. Chicago: Rand McNally.

Cooper, C. (1971). St. Fracis square: Attitudes of its residents. *A.I.A. Journal.* December, 22.

Couclelis, H., & Golledge, R.G. (1983). Analytic research, positivism, and behavioural Geography. *Annals of the Association of American Geographers,* 73, 331-339.

Cox, V., Paulus, P.B.,& McCain, G. (1984). Prison crowding research: The relevance of prison housing standards and a general approach regarding crowding phenomena. *American Psychologists,* 39, 1148-1160.

Cox, V., Paulus, P.B., McCain, G., & Schkade, W. (1979). Field research on the effects of crowding in prisons and on offshore drilling platform. In J. Aiello and A. Baum (Eds.), *Residential Crowding and Design.* New York: Plenum.

Craik, K.H. (1970).The environmental dispositions of environmental decision makers. *Annals,* 389, 80-94.

Craik, K.H. (1973). Environmental psychology. *Annual Review of Psychology,* 24, 403-422.

D'Atri, D. (1975). Psychophysiological responses to crowding. *Environment and Behavior,* 7, 237-252.

D'Atri, D., & Ostfeld, A.M. (1975). Crowding: Its effects on the elevation of blood pressure in a prison setting. *Preventive Medicine,* 4, 550-566.

Daves, W.F., & Swaffer, P.W. (1971). Effect of room size on critical interpersonal distance. *Perceptual and Motor Skills,* 6, 926.

David, A.S., & Wessely, S.C. (1995). The legend of Camelford: Medical consequences of a water pollution accident. *Journal of Psychosomatic Research,* 39, 1-9.

De Jonge, D. (1962). Images of Urban Areas: Their Structure and Psychological Foundations. *Journal of American Institute of Planners,* 28, 266-276.

Deevey, F. (1960). The hare and the haruspex. *American Scientist,* 48, 415-429.

DeLongis, A., Coyne, J., Deakof, G., Folkman, S., & Lazarus, R.S. (1982). Relationships of daily hassles, uplifts, and major life events to health status. *Health Psychology,* 1, 119-136.

Deodikar, S.G. (1991). Ecological awareness reflected in Manusmriti. In V.N. Jha (Ed.), Proceedings of the National

Seminar on Environmental Awareness, Reflected in *Sanskrit Literature* (pp. 93-100). Poona: Centre of Advance Study in Sanskrit.

Desor, J. (1972). Towards a psychological theory of crowding. *Journal of Personality and Social Psychology*, 11, 93-97.

Desor, J.A. (1972). Toward a psychological theory of crowding. *Journal of Personality and Social Psychology*, 21, 79-83.

Dharchaudhuri, I. (2003). A river dam(n)ed. *Hindustan Times*, 13th August, p. 2.

Dockery, D.W. , & Pope 111. (1994). Acute respiratory effects of particulate air pollution. *Annual Review of Public Health*, 15, 107-132.

Donnerstein, E., & Wilson, D.W. (1976). Effects of noise and perceived crowding on ongoing and subsequent aggressive behaviour. *Journal of Personality and Social Psychology*, 34, 774-781.

Dooley, B. (1974). *Crowding Stress: The Effects of Social Density on Men with Close or Far Personal Space*. Unpublished doctoral dissertation, University of California, Los Angeles.

Downs, R.M. (1970). The cognitive structure of an urban shopping centre. *Environment and Behavior*, 2, 13-39.

Downs, R.M., & Stea, D. (1977). *Maps in Minds: Reflections on Cognitive Mapping*. New York: Harper & Row.

Downs, R.M., & Stea, D. (Eds.) (1973). *Images and Environment: Cognitive Mapping and Spatial Behavior*, Chicago: Aldine.

Dubos, R. (1980). *Man Adapting*, 2nd edition. New Haven, CT: Yale University Press.

Duffy, E. (1962). *Activation and Behavior*. New York: Wiley.

Duke, M.P., & Nowicki, S. (1972). A new measure of social learning model for interpersonal distance. *Journal of Experimental Research in Personality*, 6, 119-132.

Dumont, J. (1989).Validity of multidimensional scaling in the context of structured conceptualization. Special issue. Concept mapping for evaluation and planning. *Evaluation and Program Planning*, 12, 81-86.

Dwivedi, C.B., Mishra, R.K., & Tripathi, B. (1984). Noise induced arousal in sentence processing. *Perspectives in Psychological Researches*, 7, 24-25.

Ebbesen, E.P., Kjos, G.L., & Konecni, V.J. (1976). Spatial ecology: Its effects on choice of friends and enemies. *Journal of Experimental Social Psychology*, 9, 505-518.

Edney, J.J. (1975). Territoriality and control: A field experiment. *Journal of Personality and Social Psychology*, 31, 1108-1115.

Edney, J.J., & Grundmann, M.J. (1979). Friendship, group size, and boundary size: Small group spaces. *Small Group Behavior*, 10, 124-135.

Edney, J.J., & Jorden-Edney, N.L. (1974). Territorial spacing on the beach. *Sociometry*, 37, 92-104.

Edney, J.J., Walker, C.A., & Jordan, N.L. (1976). Is there reactance in personal space? *Journal of Social Psychology*, 100, 207-217.

Ehrlich, P., Ehrlich, A. (1970). *Population Resources and Environment*. San Francisco: Freeman.

Emiley, S.F. (1975). The effects of crowding and interpersonal attraction on affective responses, tasks performance and verbal behaviour. *Journal of Social Psychology*, 97, 267-271.

EMS (April, 2003). *Air Pollution and Health Studies*. Washington Post Study.

EPA (March, 1991). *Risk Assessment for Toxic Air Pollutants: A Citizen's Guide*, 450/3-90-024.

EPA (2000). *Toxic Release Inventory 2000,* Public Data release, USA.

EPA (Feb, 2002). Associated press story on *Air Pollution and Health Studies.*

Epstein, Y., & Baum, A. (1978). Crowding: Methods of study. In A. Baum & Y. Epstein (Eds.), *Human Response to Crowding*, Hillsdale, NJ: Erlbaum.

Esser, A.H. (1974). Environment and mental health. *Science, Medicine, and Man*, 1, 181-193.

Evans, G. (1975). *Behavioral and Physiological Consequences of Crowding in Humans.* Unpublished doctoral dissertation, University of Massachusetts.

Evans, G. (1978). Human spatial behaviour: The arousal model. In A. Baum & Y. Epstein (Eds.) *Human Response to Crowding* (pp. 283-303). Hillsdale, NJ: Erlbaum.

Evans, G. (1979). Crowding and human performance. *Journal of Applied Social Psychology*, 9, 27-46.

Evans, G. (1980). Environmental cognition. *Psychological Bulletin*, 88, 259-287.

Evans, G. (1996). *Current Trends in Environmental Psychology.* IAAP Newsletter.

Evans, G., & Cohen, S. (1987). Environmental stress. In D. Stokols and I. Altman (Eds.), *Handbook of Environmental Psychology*, New York: John Wiley & Sons.

Evans, G., & Howard, R.B. (1973). Personal space. *Psychological Bulletin*, 80, 334-344.

Evans, G., & Lepore, S. (1992). Household crowding and social support: A quasi-experimental analysis. *Journal of Personality and Social Psychology*, 57, 994-999.

Evans, G., & Pezdek, K. (1980). Cognitive mapping: Knowledge of real world distance and location information. *Journal of Experimental Psychology: Human Learning and Memory*, 6, 13-24.

Evans, G., Brennan, P., Skorpanich, M.A., & Held, D. (1984). Cognitive mapping and elderly adults: Verbal and location memory for urban landmarks. *Journal of Gerontology*, 39, 452-457.

Evans, G., Cohen, S. (1987). Environmental stress. In D. Stokols and I. Altman (Eds.), *Handbook of Environmental Psychology*, New York: Wiley.

Evans, G., Jacobs, S.V., & Frager, N. (1982). Behavioral response to air pollution. In A. Baum & J. Singer (Eds.), *Advances in Environmental Psychology* (Vol. 4) pp. 237-270. Hillsdale, NJ: Erlbaum.

Evans, G., Palsane, M., Lepore, S., & Martin, J. (1989). Residential density and psychological health: The mediating effect of social support. *Journal of Personality and Social Psychology*, 57, 994-999.

Everest, L. (1986). *Behind the Poison Cloud.* Hyderabad. Banner Press.

Everett, P.B. (1977). *A behavioural Science Approach to Transportation Systems Management.* Unpublished manuscript.

Felipe, N., & Sommer, R. (1966). Invasion of personal space. *Social Problems*, 14, 206-214.

Festinger, L., Schachter, S., & Back, K. (1950). *Social Pressures in Informal Groups*. New York: Harper & Row.

Fischer, J.D., Bell, P., & Baum, A. (1984). *Environmental Psychology* (2nd ed.), New York: Holt, Rinehart & Winston.

Fisher, R.L. (1967). Social schema of normal and disturbed school children. *Journal of Educational Psychology*, 58, 88-92.

Foot, H.C., Chapman, A.J., & Smith, J.R. (1977). Friendship and social responsiveness in boys and girls. *Journal of Personality and Social Psychology*, 35, 401-411.

Forston, R.F., & Larson, C.U. (1968). The dynamics of space: An experimental study in proxemic behaviour among Latin American and North Americans, *Journal of Communication*, 18, 109-116.

Fox, S. (1991). Self and world: A transpersonal, ecological approach. *Revision*, 1, 13.

Fox, W. (1986). Approaching Deep Ecology: A response to Richard Sylvan's Critique of Deep Ecology. *Environmental Studies Occasional Paper* 20, Hobart: University of Tasmania.

Foxx, R.M., & Hake, D.F. (1977). Gasoline conservation: a procedure for managing and reducing the driving of college students. *Journal of Applied Behavioral Analysis*, 10, 61-74.

Frankenhaeuser, M., & Johansson, G. (1986). Stress at work: Psychological and Psychosocial aspects. *International Review of Applied Psychology*, 35, 287-299.

Freedman, J. (1975). *Crowding and Behavior*. San Francisco: Freeman.

Freedman, J.L., Klevansky, S., & Ehrlich, P.I. (1971). The effect of crowding on human Task performance. *Journal of Applied Social Psychology*, 1, 7-26.

Freedman, J.L., Levy, A., Buchanan, R., & Price, J. (1972). Crowding and human aggressiveness. *Journal of Experimental Social Psychology*, 8, 528-548.

Fried, M. (1963). Grieving for a lost home. In L. J. Duhl (Ed.), *The Urban Condition*. New York: Simon & Schuster.

Fry, A.M., & Willis, F.N. (1971). Invasion of personal space as a function of age of the invader. *Psychological Record*, 2, 385-389.

Fuller, B. (1957). Man with the chronofile. *Sunday Review*, April 14-18.

Fuller, T.D., Edwards, J.N., Vorakitphokatorn, S., & Sermsri, S. (1993). Household crowding and family relations in Bangkok. *Social Problems*, 40, 410-430.

Galle, O., Gove, W., & McPherson, J. (1972). Population density and pathology: What are the relationships for man? *Science*, 176, 2-30.

Galster, G. (1987). Identifying correlates of the dwelling satisfaction: An empirical Critique. *Environment and Behavior*, 19, 539-568.

Gans, H.J. (1967). *The Levittowners*. New York: Pantheon Books.

Garling, T., & Golledge, R.G. (1989). Environmental perception and cognition. In H. Zube and G.T. Moore (Eds.), *Advances in Environmental Behavior and Design* (Vol. 2, pp.203-236), New York: Plenum.

Garling, T., Book, A., Lindberg, , E., & Nilsson, T. (1981). Memory for the spatial layout of the everyday physical environment: Factors affecting the rate of acquisition. *Journal of Experimental Psychology*, 1, 263-277.

Garling, T., Lindberg, E., & Mantyla, T. (1983). Orientations in buildings: Effects of familiarity, visual access, and orientation aids. *Journal of Applied Psychology*, 68, 177-186.

Geller, E.S., Winnett, R.A., & Evert, P.B. (1982). *Preserving the Environment: New Strategies for Behavioral Change*. New York: Pergamon.

Gibson, J.J. (1966).*The Senses Considered as Perceptual Systems*. Boston: Houghton Miffin.

Gibson, J.J. (1979). *The Ecological Approach to Visual Perception*. Boston: Houghton Miffin.

Gifford, R. (1982). Projected interpersonal distance and interpersonal choices: Personality, sex, and social situation. *Social Psychology Quarterly*, 45, 145-152.

Gifford, R. (1987). *Environmental Psychology: Principles and Practice*. Boston: Allen & Bacon.

Gifford, R., Peacock, J. (1979). Crowding: More fearsome than crime provoking? Comparison of an Asian city and a North American city. *Psychologia*, 22, 79-83.

Glass, D., & Singer, J. (1972). *Urban Stress*. New York: Academic Press.

Glass, D.C., & Singer, J.E. (1972). *Urban Stress*. New York: Academic Press.

Goduka, I.N., Poole, D.A., & Aotaki-Phenice, L. (1992). A comparative study of black South African children from three different contexts. *Child Development*, 63, 509-525.

Goeckner, D., Greenough, W., & Meed, W. (1973) Deficits in learning tasks following overcrowding in rats. *Journal of Personality and Social Psychology*, 28, 256-261.

Goffman, E. (1963). *Behavior in Public Places.* New York: Free Press.

Goldsmith, J., & Friberg, L. (1977). Effects of air pollution on human health. In A. C. Stern (ed.), *Air Pollution* (3rd ed, Vol. 3, pp. 458-610). New York: Academic.

Golledge, R.G. (1977). Multidimensional analysis in the study of environmental behaviour and environmental design. In I. Altman and J.F. Wohlwill (Eds.), *Human Behavior and Environment: Advances in Theory and Research* (Vol. 2, pp. 1-42). New York: Plenum.

Golledge, R.G. (1987). Environmental Cognition. In D. Stokols & I. Altman (Eds.), *Handbook of Environmental Psychology*. (Vol. 1, pp. 131-174), New York: John Wiley & Sons.

Golledge, R.G., & Rushton, G. (1976). *Spatial Choice and Spatial Behavior*. Columbus: Ohio State University Press.

Golledge, R.G., Briggs, R., & Demko, D. (1969). The configuration of distances in intra-Urban space. *Proceedings of the Association of American Geographers,* 1, 60-65.

Golledge, R.G., Parnicky, J.J.,& Rayner, J.N. (1979). *The Spatial Competence of Selected Populations* (Vols. 1&2). Columbus: Ohio State University Research Foundation.

Golledge, R.G., Rayner, J.N., & Rivizzigno, V.L. (1982). Comparing objective and cognitive representations of environmental cues. In R.G. Golledge and J.N. Rayner (Eds.), *Proximity and Preference: Problems in the Multidimensional Analysis of Large Data Sets* (pp. 233-266). Minneapolis: University of Minnesota Press.

Gove, W., & Hughes, M. (1983). *Overcrowding in the Households.* New York: Academic Press.

Gove, W., Hudges, M., & Galle, O. (1979). Overcrowding in

the home: An empirical investigation of its possible pathological consequences. *American Sociological Review*, 44, 59-80.

Green, R.G., & O'Neal, E.C. (1969). Activation of cue-elicited aggression by general Arousal. *Journal of Personality and Social Psychology*, 11, 289-292.

Greenberg, C.I. (1979). Toward an integration of ecological psychology and industrial Psychology: Undermanning theory, organization size and job enrichment. *Environmental Psychology and Nonverbal Behavior*, 3, 228-245.

Greenberg, R.A., Strecher, V., Bauman, K., & Boat, B. (1994). Evaluation of a home based intervention programme to reduce infant passive smoking and lower respiratory illness. *Journal of Behavioural Medicine*, 17, 273-290.

Guardo, C.J., & Meisels, M. (1971). Child-parent spatial patterns under praise and Reproof. *Developmental Psychology*, 5, 365.

Gubrium, J.F. (1974). Victimization in old age. *Crime and Delinquency*, 20, 245-250.

Haber, G.M. (1980). Territorial invasion in the classroom: Invadee response. *Environment and Behavior*, 12, 17-31.

Hake, D.F., & Foxx, R.M. (1978). Promoting gasoline conservation: The effects of reinforcement schedules, a leader and self-recording. *Behavior Modification*, 2, 339-369.

Hall, C.T. (1959. *The Silent Language*. New York: Doubleday.

Hall, C.T. (1966).*The Hidden Dimension*. New York: Doubleday.

Halmiova, O., & Potasova, A. (1994). The effects of environmental neurotoxins on performance of children in sensorimotor tasks. *Studia Psychological*, 36, 103-111.

Halpern, D. (1995). *Mental Health and the Built Environment*. London: Taylor & Francis.

Halpern, D. (1995). *Mental Health and the Built Environment*. London: Taylor & Francis Hazarika, S. (1987). *Bhopal: The Lessons of a Tragedy*. New Delhi: Penguin Books.

Halpern, D.S. (1987). Architectural preferences and mere exposure effects: The role of recognition in attitudinal enhancement by exposure: Cited in D.Halpern (1995), *Mental Health and the Built Environment*. London: Taylor & Francis.

Hardin, G. (1968). The tragedy of the commons. *Science*, 162, 1243-1248.

Hauser, P., & Dunken, O. (1959). *The Study of Population: An Inventory of Appraisal. Chicago*: University of Chicago Press.

Hayduk, L.A. (1983). Personal space: Where we now stand. *Psychological Bulletin*, 94, 293-335.

Hayduk, L.A. (1985). Personal space: The conceptual and measurement implications of structural equation models. *Canadian Journal of Behavioral Science*, 17, 140-149.

Hayes, S.C., & Cone, J.D. (1977). Reducing residential electrical energy use: Payments, Information and feedback. *Journal of Applied Behavioral Analysis,* 10, 425-435.

Hayes-Roth, B. (1977). Evolution of cognitive structures and processes. *Psychological Review*, 84, 260-278.

Hayes-Roth, B. (1979). Distinguishing theories of representation: A critique of Anderson's "arguments concerning mental imagery." *Psychological Review*, 86, 376-382.

Hazen, N.L., Lockman, J.J., & Pick, H. L., Jr. (1978). The development of children's representations of large-scale environments. *Child Development,* 4, 71-115.

Hebb, D.O. (1949). *The Organization of Behavior.* New York: Wiley.

Hebb, D.O. (1972). *Textbook of Psychology* (3rd ed.). Philadelphia: Saunders.

Helson, H. (1964). *Adaptation-Level Theory: An Experimental and Systematic Approach to Behavior.* New York: Harper.

Hepper, P. (1982). Smoking, passive smoking and smell. *Medical Science Research,* 20, 265-266.

Herridge, C.F. (1974). Aircraft noise and mental health. *Journal of Psychomotor Research*, 18, 239-243.

Hershberger, R.G., & Cass, R.C. (1974). Predicting user responses to buildings. In D. Carson (Ed.), *Man-Environmental Interactions: Evaluations and Applications* (Part 2), Stroudsburg, Pa: Dowden, Hutchinson & Ross.

Hill-Lynch, K. (1960). *The Image of the City*. Cambridge, MA: MIT Press, Cambridge.

Holahan, C.J. (1986). Environmental psychology. *Annual Review of Psychology*, 37, 381-407.

Hopkins, J. (1994). Orchestrating an indoor city: Ambient noise inside a megamall. *Environment and Behaviour*, 26, 785-812.

Hoppe, R.A., Greene, M.S., & Kenny, J.W. (1972). Territorial markers: Additional findings. *Journal of Social Psychology*, 88, 305-306.

Horowitz, M.J. (1968). Spatial behaviour and psychopathology. *Journal of Nervous and Mental Disease*, 146, 24-35.

Hourihan, K. (1984). Context-dependent models of residential satisfaction. *Environment and Behavior*, 16, 369-393.

Hutt, C., & Vaizey, M.J. (1966). Differential effects of group density on social behaviour. *Nature*, 209, 1371-1372.

ICMR (1986). *Health Effects of Exposure to Toxic Gas at Bhopal.* New Delhi.

Ingram, R.E., & Geller, E. S. (1975). A community integrated behaviour modification approach to facilitating paper recycling. *JSAS Catalog of Selected Documents in Psychology.* 5, 327.

Ittelson, W.H. (1951). The consistency in perceptual theory. *Psychological Review*, 58, 285-294.

Ittelson, W.H., Proshansky, H., Rivlin, L.G., & Winkel, G.H. (1974). *An Introduction to Environmental Psychology.* New York: Holt, Rinehart & Winston.

Jacobs, J. (1961). *The Death and Life of Great American Cities.* New York: Random House.

Jain, U. (1976). *Competition Tolerance and Need Hierarchy Patterns as Related to Population Growth.* Project Report of the ICSSR.

Jain, U. (1978). Competition tolerance in high-low density in urban-rural areas. *Journal of Social Psychology*, 105, 297-298.

Jain, U. (1987). *Psychological Consequences of Crowding.* New Delhi: Sage.

Jain, U. (1997). *The feeling of crowding and its psychological consequences.* New Delhi: Har-Anand Publications.

Jain, U. (Ed). (1997). Perspectives on Environment and Behavior. New Delhi: Har-Anand. *Project Report*, New Delhi: ICSSR.

Jain, U., & Palsane, M.N. (2004). Environment and behaviour. In J. Pandey (Ed), Psychology in Indian Revisited: Developments in the Discipline (Vol. 3): Applied *Social and Organizational Psychology* (pp. 261-308). New Delhi: Sage.

Jain, U., & Palsane, M.N. (2004). ICSSR survey in Psychology, Vol. 3, New Delhi: Sage.

Jain, U., & Suraj, M. (1983). Effects of prolonged deprivation on attribution of causes of success and failure. *Journal of Social Psychology*, 124, 143-49.

Jarvis, M.J., Foulds, J., & Feyerabend, C. (1992). Exposure to passive smoking among bar staff. *British Journal of Addiction*, 87, 111-113.

Jha, V.N. (Ed.) (1991). *Proceedings of the National Seminar on Environmental Awareness Reflected in Sanskrit Literature.* Poona: Centre of Advance Study in Sanskrit.

Johnson, D., & Matross, R. (1977). Interpersonal influence in psychotherapy: A Socio-Psychological view. In A. Gurman & A. Razin (Eds.), *Effective psychotherapy: A Handbook of Research.* New York: Pergmon.

Jones, S.E., & Aiello, J.R. (1973). Proximic behaviour of black and white first, third, and fifth-grade children. *Journal of Personality and Social Psychology*, 25, 21-27.

Jorgenson, D.O., & Dukes, F.O. (1976). Deindividuation as a function of density and Group membership. *Journal of Personality and Social Psychology*, 34, 24-39.

Kahneman, D. (1973). *Attention and Effort.* Englewood Cliffs, NJ: Prentice-Hall.

Kaplan, K.J. (1977). Structure and process in interpersonal "distancing". *Environmental Psychology and Nonverbal Behavior*, 1, 104-121.

Kaplan, R. (1974). Some methods and strategies in the prediction of preference. In E.H. Zube, J.G. Fabos, & R.O. Brush (Eds.), *Landscape Assessment, Values, Perceptions and Resources.* Stroudsburg, Pa: Dowden, Hutchinson & Ross.

Kaplan, R. (1976). Way-finding in the natural environment. In G. T. Moore & R.G. Golledge (Eds.), *Environmental Knowing* (pp. 46-57). Stroudsburg, PA: Dowden, Hutchinson & Ross.

Kaplan, S. (1973). Cognitive maps in perception and thought. In R. Downs & D. Stea (Eds.), *Image and Environment: Cognitive Mapping and Spatial Behaviour*, Chicago: Aldine.

Kaplan, S. (1975). An informal model for the prediction of preference. In E.H. Zube, R.O. Brush., & J.G. Fabos (Eds.), *Landscape Assessment.* Stroudsburg, PSW-35. Berkeley, CA,

Kaplan, S. (1979). Perception and landscape: Conceptions and Misconceptions. In G. Elsner & R. Smardon, *USDA Forest Service Report PSW-35*, Berkeley, CA.

Kaplan, S., & Kaplan, R. (1978). *Humanscape: Environments for people.* North Scituate, MA: Duxbury.

Kaplan, S., & Kaplan, R. (1982). *Cognition and Environment: Functioning in an Uncertain World.* New York: Praeger.

Karlin, R., Epstein, Y., & Aiello, J. (1978). *Strategies for investigation of crowding.* In A. Esser., & B. Greenberg (Eds.), designs for community and privacy, New York: Plenum Press.

Kasmer, A. (1976). The development of a useful lexicon of environmental descriptors. *Environment and Behavior,* 2, 135-169.

Kates, R.W., & Wohlwill, J.F.(Eds.) (1966). Man's response to the physical environment. *Journal of Social Issues,* 22, 1-140.

Katz, D. (1937). *Animals and Men.* New York: Longmans, Green.

King, M.G. (1966). Interpersonal relations in preschool children and average approach distance. *Journal of Genetic Psychology,* 109, 109-116.

Kleck, R. (1969). Physical stigma and task oriented interaction. *Human Relations,* 22, 53-60.

Klein, K., & Harris, B. (1979). Disruptive effects of disconfirmed expectancies about crowding. *Journal of Personality and Social psychology,* 37, 391-395.

Klopfer, P.H. (1968). From Ardrey to altruism: A discourse on the biological basis of human behaviour. *Behavioral Science,* 13, 399-401.

Knipschild, P. (1977). Medical effects of aircraft noise. *International Archives of Occupational and Environmental Health,* 40, 185-204.

Knowles, E.S. (1980). An affiliative conflict theory of personal and group spatial behaviour. In P.B. Paulus (Ed.), *Psychology of Group Influence.* Hillsdale, NJ: Erlbaum.

Konecni, V. (1975). The mediation of aggressive behaviour: Arousal level vs. anger and Cognitive labeling. *Journal of Personality and Social Psychology,* 32, 706-712.

Konecni, V.J., Libuser, L., Motron, H., & Ebbesen, E.B. (1975). Effects of a violation of personal space on escape and helping responses. *Journal of Experimental Social Psychology,* 11, 288-299.

Korte, C. (1979). Helpfulness in urban environment. In A. Baum., J.E. Singer, & S. Valins (Eds.), *Advances in Environmental Psychology: The Urban Environment* (Vol 1). Hillsdale, NJ: Erlbaum.

Krupat, E. (1987). *People in Cities: The Urban Environment and its Effects*. New York: Cambridge University Press.

Kryter, K.D. (1970). *The Effects of Noise on Man*. New York: Academic.

Kuethe, J.L. (1962). Social schemas. *Journal of Abnormal and Social Psychology*, 64, 31-38.

Kuipers, B. (1980). A model of the acquisition of spatial knowledge. In L. Bowl (Ed.), *Models of Dialogue: Theory and Application*. London: McMillan.

Kuipers, B. (1982). The 'map in the head' metaphor. *Environment and Behavior*, 4, 202-220.

Kulkarni, K. (1984). *Geography of Crowding and Human Response*. New Delhi: Concept.

Kulkarni, N. (1991). Man and nature relationship reflected in the Sukla Yajurved. In V.N. Jha (Ed.), *Proceedings of the National Seminar on Environmental Awareness Reflected in Sanskrit Literature* (pp. 1-6). Poona Centre of Advance Study in Sanskrit.

Kumar, P., & Mathur, C.N. (1969). Neuroticism as a factor in noise distractibility. *Journal of Indian Academy of Applied Psychology*, 6, 27.

Kumar, P., & Singh, J. (1986). Deteriorating urban environment: A case of Calcutta. In K.K. Dubey & A.K. Singh (eds.), *Urban Environment in India* (pp. 389-401), New Delhi: Inter India Publications.

Kuper, L. (1953). Blueprint for living together. In L. Kuper (Ed), *Living in Towns* (pp. 1-202). London: Cresset Press.

Langer, E.J., & Saegert, S. (1977). Crowding and cognitive control. *Journal of Personality and Social Psychology*, 35, 175-182.

Latane, B., & Darley, J. M. (1970). *The Unresponsive Bystander: Why Doesn't he Help?* New York: Prentice-Hall.

Lawrence, (1974). Science and sentiment: Overview of research on crowding and human Behavior. *Psychological Bulletin*, 712-720.

Layhausen, P. (1965). The same community: A density problem? *Discovery*, 26, 27-33.

Lazarus, R.S. (1966). *Psychological Stress and the Coping Process*. New York: McGraw-Hill.

Lazarus, R.S., & Cohen, J. (1977). Environmental stress. In J. Wohlwill & I. Altman (Eds.), *Human Behavior and Environment*. (pp. 90-127). New York: Plenum.

Lazarus, R.S., & Folkman, S. (1984). *Stress, Appraisal and Coping*. New York: Springer. Morin, M. (1989). Applied social psychology and noise in work situations. *Cahiers Internationaux de Psychologie Sociale*, March, 1, 31-51.

Lee, T. (1976). *Psychology and the Environment*. London: Methuen.

Left, H.I. (1974). *Experience, Environment and Human Potentials*. New York: Oxford University Press.

Lepore, D., Merritt, k., Kawasake, N., & Mancuso, R. (1990). *Social Withdrawal in Crowded Residences*. Paper presented at the meeting of the Western Psychological Association, Los Angeles, CA.

Lepore, S., Evans, G., & Schneider, M. (1991). Dynamic role of social support in the link Between chronic stress and psychological distress. *Journal of Personality and Social Psychology*, 61, 899-909.

Lestor, D. (1970). Social disorganization and completed suicide. *Social Psychiatry*, 5, 175-176.

Levy, L., & Herzog, A. (1974). Effects of population density and crowding on health and social adaptation in the Netherlands. *Journal of Health and Social Behavior*, 15, 240.

Lewin, K. (1936). *Principles of Topological Psychology*. New York: McGraw-Hill.

Lewin, K. (1951). *Field Theory in Social Science*. New York: Harper & Row.

Ley, D., & Cybriwsky, R. (1974). The spatial ecology of stripped cars. *Environment and Behavior*, 6, 53-68.

Lingwood, D.A. (1971). Environmental education through information seeking: The case of an environmental teach in. *Environment and Behavior*, 3, 220-262.

Little, K.B. (1965). Personal space. *Journal of Experimental Social Psychology*, 1, 237-247.

Loo, C. (1973). Important issues in researching the effects of crowding on humans. *Representative Research in Psychology*, 26, 27-33.

Lorenz, K. (1966). *On Aggression.* New York: Harcourt Brace Jovanovich.

Lorenz, K. (1970). *On Aggression.* London: Methuen.

Loring, W. (1955). Housing characteristics and social disorganization. *Social Problems*, 3, 160-168.

Lott, B.S., & Sommer, R. (1967). Seating arrangements and status. *Journal of Personality and Social Psychology*, 7, 90-95.

Love, K.D., & Aiello, J.R. (1980). Using projective techniques to measure interaction distance: A methodological note. *Personality and Social Psychology Bulletin*, 6, 102-104.

Lovelock, J. (1979). *Gaia: A New Look at Life on Earth.* New York: Oxford University Press.

Lundberg, U. (1976). Urban commuting: Crowdedness and catecholamine excretion. *Journal of Human Stress*, 2, 26-32.

Lynch, K. (1960). *The Image of the City.* Cambridge, MA: MIT Press. Cambridge.

MacDonald, W., & Oden, C. (1973). *Effects of Extreme Crowding on the Performance of Five Married Couples During Twelve Weeks of Intensive Training.* Proceedings of the 81st Annual Convention of the American Psychological Association, 8, 209-210.

Mandler, J. (1995). Leaving the Earth Space Colonies, Disney and Epcot. In G. Sessions (Eds.) *Deep Ecology for the 21st century*, Massachusetts: Shambhala Publications.

Mardson, H. (1972). Crowding and population density. In J.F. Wohlwill & D.H. Carson (Eds), *Environment and the Social Sciences: Perspectives and Applications*, APA, Washington, D.C.

Marsella, A., Escudero, M., & Gorden, P. (1970). The effects of dwelling density on mental disorders in Felipino men. *Journal of Personality and Social Psychology*, 11, 288-294.

Mason, J. W. (1968). A historical review of the stress field. *Journal of Human Stress*, 1, 6-12.

Maurer, R., & Baxter, J. (1972). Images of neighbourhood among black, Anglo and Mexican-American children. *Environment and Behavior*, 4, 351-388.

Mawby, R. (1977). Defensible space: A theoretical and empirical appraisal. *Urban Studies*.14, 169-179.

Mazur, A. (1977). Interpersonal spacing on public benches in "contact" vs. "non-contact" cultures. *Journal of Social Psychology*, 101, 53-58.

McAndrew, E.T., Ryckman, R.M., Horr, W., & Solomon, R. (1978). The effects of invader placement of spatial markers on territorial behaviour in a college population. *Journal of Social Psychology*, 104, 149-150.

McBride, G., King, M.G., & James, J.W. (1965). Social proximity effects on galvanic skin response in adult humans. *Journal of Psychology*, 61, 153-157.

McBurney, D.H. (2002). *Research Methods*. Singapore: Thomson Asia Pvt. Ltd.

McCain, G., Cox, V., & Paulus, P.B. (1976). The relationship between illnesses complains and degree of crowding in a prison environment. *Environment and Behavior*, 8, 289-291.

McCarthy, D., & Saegert, D. (1978). Residential density, social overload, and social withdrawal. *Human Ecology*, 6, 253-272.

McCauley, C., & Newman, J. (1977). Eye contact with strangers in a city, suburb, and small town. *Environment and Behaviour*, 9, 547-558.

McCauley, C., Coleman, G., & DeFusco, P. (1977). *Commuters' Eye Contact with* Strangers in City and Suburban Train Stations. Evidence of Short-Term Adaptation to *Interpersonal Overload in the City*. Unpublished manuscript, Bryn Mawr college, Bryn Mawr, PA.

McKechnie, G.E. (1974). *ERI Manual: Environmental Response Inventory*. Berkeley, CA: Consulting Psychologists Press.

McLaughlin, A. (1995). The heart of deep ecology. In George Sessions (Ed.), *Deep Ecology for the 21st Century* (pp. 85-94).

Mehrabian, A. (1969). Significance of posture and position in the communication of attitude and status relationships. *Psychological Bulletin*, 71, 359-372.

Mehrabian, A., & Diamond, S.G. (1971). Effects of furniture arrangement, props, and personality on social interaction. *Journal of Personality and Social Psychology*, 20, 18-30.

Mehrabrian, A., & Russell, J.A. (1974). *An Approach to Environmental Psychology*. Cambridge, MA: MIT Press.

Merry, S. (1981). Defensible space undefended: Social factors in criminal control through environmental design. *Urban Affairs Quarterly*, 16, 397-422.

Merton, R.K. (1948). The social psychology of housing. In W. Dennis (Ed.), *Current Trends in Social Psychology* (pp. 163-217). Pittsburgh, PA: University of Pittsburg Press.

Meyers, A.E., Artz, L.M., Craighead, W.E. (1976). The effects of instructions, incentives, and feedback on a community problem: Dormitory noise. *Journal of Applied Behavioral Analysis*, 9, 445-457.

Meyers, N. (1984). *Gaia: An Atlas of Planet Management*. Garden City, N.Y.: Anchor Press.

Michelson, W. (1970). *Man and his urban environment: A sociological approach.* Reading, MA: Addison-Wesley.

Michelson, W., & Reed, P. (1975). A time budget record sheet. In W. Michelson (Ed.), *Behavioral Research Methods in Environmental Design.* Stroudsburg, Pa: Dowden, Hutchinson, & Ross.

Michelson, W.H. (1977). *Environmental Choice, Human Behavior, and Residential Satisfaction.* New York: Oxford University Press.

Middlemist, R.D., Knowles, E.S., & Matter, C.F. (1976). Personal space invasion in the lavatory: Suggestive evidence for arousal. *Journal of Personality and Social Psychology*, 33, 541-546.

Milgram, S, (1970). *The experience of living in big cities. Science,* 167, 1461-1468.

Milgram, S. (1970). The experience of living in cities. *Science,* 167, 1461-1468.

Milgram, S. (1970). The experience of living in cities. *Science,* 167, 1461-1468.

Milgram, S., Greenwald, J., Keesler, W., McKenna, W., & Waters, J. (1972). A psychological map of New York city. *American Scientist*, 60, 194-200.

Miller, J.G. (1964). *Adjusting to Overloads of Information. Disorders of Communication,* Proceedings of the Conference of the Association for Research in Nervous and Mental Disease, December 7-8, 1962. New York.

Miller, N. (1944). Experimental studies in conflict. J. McV. Hunt (Ed.), *Personality and the Behavior Disorders* (Vol. 1). New York: Ronald.

Minski, M.A. (1975). A framework for representing knowledge. In P. H. Winston (Ed.), *The Psychology of Computer Vision.* New York: McGraw- Hill.

Mishra, A.B. (1987). Mining a hill and undermining a society: The case of Gandhamardan. In A. Agarwal, D.' Minte & U. Samarth (Eds.), *The Fight for survival: Peoples' Action for Environment,* pp. 125-144.

Misra, G. (1988) *Psycho-Social Competence in Children in MIC Exposed Areas of Bhopal.* Paper presented at the workshop on developmental norms of children,

Mitchell, R. (1971). Some implications of high density housing. *American Sociological Review*, 36, 1461-1468.

Mitra, S.K. (Ed.) (1972). *A Survey of Research in Psychology.* Bombay: Popular Prakashan.

Moore, G.T. (1979). Knowing about environmental knowing: The current state of theory and research on environmental cognition. *Environment and Behavior*, 11, 33-70.

Moore, G.T., & Golledge, R.G. (1976). *Environmental Knowing: Theories, Research and Methods.* Stroudsburg, PA: Dowden, Hutchinson & Ross.

Moos, R.H., & Gerst, M.S. (1974). *University Residence Environment Scale.* Palo Alto, Calif: Consulting Psychologists Press.

Moos, R.H., & Lemke, S. (1984). Supportive residential setting for older people. In I. Altman, M.P. Lawton, & J.F. Wohlwill (Eds.), *Elderly People and the Environment: Vol 7. Human Behaviour and Environment: Advances in Theory and Research* (pp. 159-190).New York: Plenum.

Morgan, J.J. (1916). The overcoming of distraction and other resistances. *Archives of Psychology*, 35, 24, 1-84.

Moulik, T.R. (1981). Psychology of Poverty in India. In Udai Pareek (Ed.), *A survey of research in psychology*, 1971-76 (Part 11, pp. 655-681), Bombay: Popular Prakashan.

Murphy-Berman, V., & Berman, J. (1978). The importance of choice and sex in invasions of interpersonal space. *Personality and Social Psychology Bulletin*, 4, 424-428.

Naess, A. (1995). The shallow and the deep, long-range ecological movements. In G. Sessions (Ed.) *Deep Ecology for the 21st Century*, pp. 151-155.

Nagar, D. (1985). *Experiences and consequences of crowding.* Unpublished doctoral dissertation, Allahabad University, Allahabad.

Nagar, D. (1998). *Human Reactions to Crowding.* Jaipur: Printwell Publishers.

Nagar, D. (1999). Environmental stress and human behaviour. In G. Misra (Ed.) *Psychological Perspectives on Stress and Health.* New Delhi: Concept Publishing Company.

Nagar, D. (2005). Human rights of poor living in slums: A psycho-social analysis. In S.N. Chaudhary (Ed.), *Human Right and Poverty in India: Theoretical Issues and Empirical Evidence* (Vol. 5). New Delhi: Concept.

Nagar, D., & Mishra, M. (2001). Slum rehabilitation in Bhopal: Its psychological consequences. *Indian Cooperative Review*, 38, 171-183

Nagar, D., & Pandey, J. (1987). Affect and performance of cognitive task as a function of Crowding and noise. *Journal of Applied Social Psychology*, 17, 147-157.

Nagar, D., & Paulus, P.B. (1997). Residential crowding experience scale: Assessment and validation. *Community and Applied Social Psychology*, 7, 303-319.

Nagar, D., & Sharma, A. (2005). Influence of community living and environmental stressors on health and well-being. In A. Dalal (Ed.), *Social Dimensions of Health and Well-being.* Jaipur: Rawat Publications.

Nagar, D., Pandey, J., & Paulus, P. (1988).The effects of residential crowding experience On reactivity to laboratory crowding and noise. *Journal of Applied Social Psychology*, 18, 1423-1442.

National Institute of Public Cooperation and Child Development, New Delhi.

Newell, A., & Simon, H.A. (1972). *Human Problem Solving.* Englewood Cliffs, NJ: Prentice-Hall.

Newman, O. (1972). *Defensible Space.* New York: Macmillan.

Newman, O., & Franck, K. A. (1980). *Factor Influencing Crime and Instability in Urban Housing Developments.* Washington, D.C. National Institute of Justice.

Noesjirwan, J. (1978). A laboratory study of proxemic patterns of Indonesians and Australians. *British Journal of Social and Clinical Psychology*, 17, 333-334.

Odera, P., & Hasan, Q. (1993). Assessment of difficulties of foreign students in India. *Journal of Personality and Clinical Studies*, 9, 25-30.

Olsen, M.E. (1981). Consumers' attitudes toward energy conservation. *Journal of Social Issue*, 37, 108-131.

O'Neal, E.C., Brunault, M.A., Carifio, M.S., Tronturne, R., & Epstein, J. (1980). Effect of insult upon personal space preferences. *Journal of Nonverbal Behavior*, 5, 56-62.

Orleans, P. (1973). Differential cognition of urban residents: Effects of social scale on mapping. In R.M. Downs & D. Stea (Eds.), *Images and Environment: Cognitive Mapping and Spatial Behavior* (pp. 115-130). Chicago: Aldine.

Osgood, C., Suci, G., & Tannenbaum, P. (1957). *The Measurement of Meaning*. Urbana University of Illinois Press.

Osmond, H. (1957). Function as a basis of psychiatric ward design. *Mental Hospitals*, 8, 23-30, 83, 235-245.

Oxley, D., & Barrera, M. (1984). Undermanning theory and the workplace: Implications of setting size for job satisfaction and social support. *Environment and Behavior*, 16, 211-234.

Padaki, V., & Vyasulu, V. (1981). Towards environmental psychology. In Udai Pareek (Ed.), *A Survey of Research in Psychology*, 1971-76 (Part 11, pp. 618-654). Bombay Popular Prakashan.

Padmavati (1991). *Coping with Crowding Stress*. Unpublished doctoral dissertation, Barkatullah University, Bhopal.

Page, R.A. (1977).Noise and helping behaviour. *Environment and Behaviour*, 9, 559-572.

Palinkas, L.A., Petterson, J.S., Russell, J., & Dows, M.A. (1993). Community patterns of Psychiatric disorders after the Exxon Valdez oil spill. *American Journal of Psychiatry*, 150, 1517-1523.

Pandey, J. (1978). Effects of crowding on human social behaviour. *Journal of Social and Economic Studies*, 6, 85-95.

Pandey, J. (2004). *Psychology in Indian Revisited: Developments in the Discipline (Vol. 1-3): Applied Social and Organizational Psychology*. New Delhi: Sage.

Pandey, J., & Nagar, (1988). Residential crowding in India: Its social psychological consequences. Paper presented at the 24th Inter-national Congress of Psychology, Sydney.

Pandey, J., & Nagar, D. (1987). Experiences of indwelling and outdwelling crowding and social psychological consequences. *Social Science International,* 3, 1-9.

Pareek, U. (Ed.) (1981). *A Survey of Research in Psychology,* 1971-76 (Part 1, 11), Bombay: popular Prakashan.

Pathak, A.M., & Vohra, S.S. (1997). Environmental pollution: Awareness, attitude and behavioural concomitant among youth. In U. Jain (ed.), *Perspectives on Environment and Behaviour,* pp. 138-161, New Delhi: Har-Anand Publications.

Patterson, M.L. (1975). Personal space: Time to burst the bubble? *Man-Environment Systems,* 5, 67.

Patterson, M.L., Mullens, S., & Romano, J. (1971). Compensatory reactions to spatial intrusion. *Sociometry,* 34, 114-121.

Paulus, P.B. (1987). *Prison Crowding: A Psychological Perspective.* New York: Springer-Verlag.

Paulus, P.B., & Nagar, D. (1987). Environmental influences on social interaction and group development. In C. Hendrick (Eds.). *Review of Personality and Social Psychology,* 9, 68-90, Beverly Hills: Sage.

Paulus, P.B., & Nagar, D. (1989). Environmental influences in groups. In P.B. Paulus (Ed.), *Psychology of Group Influences,* Hillsdale, NY; Erlbaum.

Paulus, P.B., Annis, A. B., Seta, J.J., Schkade, J.K., & Matthews, R.W. (1976). Crowding does affect task performance. *Journal of Personality and Social Psychology,* 34, 248-253.

Paulus, P.B., McCain, G., & Cox, V. (1978). Death rates, psychiatric commitments, blood pressure, and perceived crowding as a function of institutional crowding. *Environment and Behavior,* 3, 107-116.

Paulus, P.B., McCain, G., & Cox, V. (1981). Crowding in jails. *Basic and Applied Social Psychology,* 4, 89-107.

Paulus, P.B., Nagar, D., & Camacho, M. (1991). Environmental and psychological Factors in reactions to apartments and mobile homes. *Journal of Environmental Psychology.* 11, 143-161.

Paulus, P.B., Nagar, D., Larey, T., & Camacho, M. (1996). Environmental, lifestyle, and Psychological factors in the health and well-being of military families. *Journal of Applied Social Psychology*, 26, 2053-2075.

Peake, P., & Leonard, J.A. (1971). The use of heart rate as an index of stress in blind pedestrians. *Ergonomics*, 14, 189-204.

Perlman, J.E. (1976). Grassrooting the system. *Social Policy*, 7, 4-20.

Piaget, J., & Inhelder, B. (1967). *The Child's Conception of Space.* New York: Norton.

Piaget, J., Inhelder, B., & Czeminska, A. (1960). *The Child's Conception of Geometry*, New York: Basic.

Pick, H.L, Jr., & Lockman, J.J. (1981). From frames of reference to spatial representations. In L.S. Liben, A.H. Patterson & N. Newcombe. *Spatial Representation and Behavior Across the Lifespan: Theory and Application.* New York: Academic.

Porteous, J.D. (1977). *Environment and Behavior; Planning and Everyday Urban Life.* Reading, mass: Addison-Wesley.

Potasova, A. (1992). Serial position effect: Indicator of environmental-toxic injury? *Studia Psychologia*, 34, 127-135.

Potasova, A., & Arochova, O. (1994). Effects of toxic agents in the environment of Mental functions in children: Some methodological approaches and findings. *Ceskoslovensha Psychologie*, 38, 131-142.

Powell, L. (1990). Factors associated with the under-representation of African-American In mathematics and science. *Journal of Negro Education*, 59, 292-298.

Preet, K., & Jain, U. (1984). *Effects of Population Density on Mental Health.* Paper presented at Indian Psychiatric Conference, Udaipur.

Proshansky, H., Ittelson, W., & Rivlin, L. (1970). Freedom of choice and behaviour in a physical setting. In H. Proshansky., W. Ittelson, & L. Rivlin (Eds.), *Environmental Psychology: Man and his Physical Setting* (pp. 173-182). New York: Holt, Rinehart & Winston.

Proshansky, H.M. (1972). Methodology in environmental psychology: Problems and Issues. *Human Factors*, 14, 451-460.

Proshansky, H.M. (1978). The city and self-identity. *Environment and Behavior*, 10, 147-169.

Proshansky, H.M., Ittelson, W.H., & Rivlin, L.G. (1970). *Environmental Psychology: Man and his Physical Setting*. New York: Holt, Rinchart and Winston.

Proshansky, H.M., Ittelson, W.H., & Rivlin, L.G. (1976). *Environmental Psychology* (2nd ed.). New York: Holt, Rinehart & Winston.

Raab, G.M., Thompson, G.O., Boyd, I., & Fulton, M. (1990). Blood lead levels, reaction time, inspection time and ability in Edinburgh children. *British Journal of Developmental Psychology*, 8, 101-118.

Rainwater, L. (1966). Fear and house-as-haven in the lower class. *Journal of American Institute of Planners*, 32, 23-31.

Raja, S., & Ganguli, T. (1983). Impact of exposure to noise on hearing acuity of employees in a heavy engineering industry, *Indian Journal of Medical Research*, 78, 100-113.

Rankin, R.E. (1969). Air pollution, control and public apathy. *Journal of the Air Pollution Control Association*, 19, 565-569.

Rao, T.V. (1981). Psychology of population and family planning. In Udai Pareek (Ed.), *A Survey of Research in Psychology*, (Part 11, pp. 714-746). Bombay: Popular Prakashan.

Rapoport, A. (1977). *Human Aspects of Urban Form*. New York: Pargamon.

Rapoport, A. (1980). Environmental preference, habitat selection and urban housing. *Journal of Social Issues*, 36, 118-134.

Rapoport, A. (1980). *Human Aspects of Urban Form: Towards a Man-Environment Approach to Urban Form and Design*. Elmsford, NJ: Pergamon.

Rapoport, A. (1985). Thinking about home environment: A Conceptual Framework. In I. Altman and C.M. Werner (Eds.), *Home environment* (pp. 255-286). New York: Plenum Press.

Reddy, D., Baum, A., Fleming, R., & Aiello, J. (1981). Mediation of social density by Coalition formation. *Journal of Applied Social Psychology*, 11, 529-537.

Reichel, D.A., & Geller, E.S. (1980). *Group versus Individual Contingencies to Conserve Transportation Energy*. Paper presented at the Southwestern Psychological Association meeting, Washington, D.C.

Reid, E., & Novak, P. (1975). Personal space: An unobtrusive measures study. *Bulletin of Psychonomic Society*, 5, 265-266.

Reilly, F.E. (1970). *Charles Peirce's Theory of Scientific Method.* New York: Fordham University Press.

Richardson, G.D. (1982). *Spatial Cognition*. Doctoral dissertation, University of California, Santa Barbara.

Rivers, W.H.R. (1905). Observations on the senses of the Todas. *British Journal of Psychology*, 1, 321-396.

Rodin, J. (1976). Crowding, perceived control, and response to controllable and uncontrollable outcomes. *Journal of Experimental Social Psychology*, 12, 133-158.

Rodin, J. (1976). Density, perceived choice and response to controllable and uncontrollable outcomes. *Journal of Experimental Social Psychology*, 12, 564-578.

Rohe, W. (1985). Urban planning and mental health. *Prevention in Home Sciences*, 4, 79-110.

Rohe, W., & Patterson, A.H. (1974). *The Effects of Varied Levels of Resources and Density on Behavior in a Day Care Center*. Paper presented by the Environmental Design Research Association, Milwaukee, WI.

Romney, A.K., & D'Andrade, R.G. (1964). Transcultural studies in cognition. *American Anthropologists*, 66, 140-170.

Rosenthal, R., & Rosnow, R.L. (1984). *Essentials of Behavioral Research: Methods and Data Analysis*. New York: McGraw-Hill Book Company.

Rotton, J. (1983). Affective and cognitive consequences of malodorous pollution. *Basic and Applied Social Psychology*, 4, 171-191.

Rowles, G., & Ohta, R. (1983). *Aging and Milieux*. New York: Academic.

Ruback, B., & Pandey, J. (1991). Crowding, perceived control, and relative power: An analysis of households in India. *Journal of Applied Social Psychology*, 21, 345-374.

Ruback, R., & Innes, C. (1988). The relevance and irrelevance of psychological research. The example of prison crowding. *American Psychologists*, 683-693.

Ruback, R.B., & Pandey, J. (1992). Very hot and really crowded: Quasi-experimental investigations of Indian "Tempos". *Environment and Behavior*, 24, 527-554.

Ruback, R.B., & Pandey, J. (1993). *Gender Differences in Reaction to Household Crowding: Affiliation and Stress in Rural India*. Paper presented at the Annual convention of National Academy of Psychology, Delhi University, Delhi.

Rumsey, N., Bull, R., & Gahngan, D. (1982). The effects of facial disfigurement on the proxemic behaviour of the general public. *Journal of Applied Social Psychology*, 12, 137-150.

Russell, J.A., & Ward, L.M. (1982). Environmental psychology. *Annual Review of Psychology*, 33, 651-682.

Sadik, N. (1992). *The State of World's Population*. New York: United Nations Population Fund.

Saegert, S. (1973). Crowding, cognitive overload and behavioural constraint. In W. Preiser (Ed.), *Environmental Design Research* (Vol 2, pp. 254-261). Stroudsberg, PA: Dowden, Hutchinson & Ross.

Saegert, S. (1974). *Effects of Spatial and Social Density on Arousal, Mood and Orientation*. Unpublished doctoral dissertation, University of Michigan, Ann Arbor.

Saegert, S. (1978). High density environments: their personal and social consequences. In A. Baum & Y. Epstein (Eds), *Human Responses to Crowding*, (pp. 257-281), Hillsdale, NJ: Erlbaum.

Saegert, S., & Winkel, G. (1990). Environmental Psychology. *Annual Review of Psychology*, 41, 441-477.

Sainsbury, P. (1955). *Suicide in London*. London: Chapman and Hall.

Sandhu, R.S. (1987). Not all slums are alike: A comparison of squatter housing in Delhi and Amritsar. *Environment and Behavior*, 19, 398-406.

Sarason, S.B. (1974). *The Psychological Sense of Community: Prospects for a Community Psychology*. San Francisco: Jossey-Bass.

Schachter, S. (1959). *The Psychology of Affliction*. Stanford, CA: Stanford University Press.

Schank, R., & Abelson, R. (1977). *Script, Plans, Goals, and Understanding*. Hillsdale, NJ: Erlbaum.

Schmidt, C.W., & Ulrich, R.E. (1969). Effects of group contingent events upon classroom noise. *Journal of Applied Behavioral Analysis*, 2, 171-179.

Schmidt, D., & Keating, J. (1979). Human crowding and personal control: An integration of the research. *Psychological Bulletin*, 86, 680-700.

Schmitt, R. (1963). Implications of density in Hong Kong. *Journal of American Institute of Planners*, 29, 210-217.

Schmitt, R. (1966). Density, health and social disorganization. *Journal of American Institute of Planners*, 32, 38-40.

Schneider, F.W., & Hansvick, C.L. (1977). Gaze and distance as a function of changes in interpersonal gaze. *Social Behavior and Personality*, 5, 49-53.

Schwab, J., Nadeau, S., & Warheit, G. (1979). Crowding and mental health. *Pavlovian Journal of Biological Science*, 14, 226-233.

Scott, W.E. (1966). Activation theory and task design. *Organizational Behavior and Human Performance*, 1, 3-30.

Seligman, C., Kriss, M., Darley, J.M., Fazio, R.H., Becker, L.J., & Pryor, J.B. (1979). Predicting summer energy consumption from home owners' attitudes. *Journal of Applied Social Psychology*, 9, 70-90.

Seligman, M.E. (1975). *Helplessness*. San Francisco: Freeman.

Sessions, G. (1995). *Deep Ecology for the Twenty-First Century*. Boston: Shanbhala.

Seta, J.J., Paulus, P.B., & Schkade, J.K. (1976). Effects of group size on proximity under cooperative and competitive conditions. *Journal of Personality and Social Psychology*, 34, 47-53.

Sethi, B., Sharma, M., Trivedi, J., & Singh, H. (1987). Psychiatric morbidity in patients attending clinics in gas affected areas in Bhopal. *Indian Journal of Medical Research*, 86, 45.

Severy, L., Forsyth, D.R., & Wagner, P.J. (1979). A multi-method assessment of personal space development in female and male, black and white children. *Journal of Non-verbal Behavior*, 4, 68-86.

Seyle, H. (1956). *The Stress of Life*. New York: McGraw Hill.

Shaffer, D.R., & Sadowski, C. (1975). This table is mine: Respect for marked bar room tables as a function of gender of spatial marker and desirability of locale. *Sociometry*, 38, 408-419.

Shaver, P., & Buhrmester, D. (1983). Loneliness, sex role orientation and group life: A Social needs perspective. In P.B. Paulus (Ed.), *Basic Group Processes* (pp. 259-288). New York: Springer-Verlag.

Sherrod, D. (1974). Crowding, perceived control and behavioural aftereffects. *Journal of Applied Social Psychology*, 4, 171-186.

Sherrod, D., Hage, J.N., Halpern, P., & Moore, B. (1977). Effects of personal causation And perceived control in response to an aversive environment: The more control the Better. *Journal of Experimental Social Psychology*, 13, 14-27.

Sherrod, D.R. (1974). Crowding, perceived control and behavioural aftereffects. *Journal of Applied Social Psychology*, 4, 171-186.

Siegal, A.W., & Schadler, M. (1977). Young children's cognitive maps of their classrooms. *Child Development*, 48, 388-394.

Siegal, A.W., & White, S.H. (1975). The development of spatial representations of Large-scale environments. In H.W. Reese (Ed.), *Advances in Child Development and Behavior* (Vol. 10), New York: Academic.

Singer, J.E., Lundberg, U., & Frankenhaeuser, M. (1978). Stress on the train: A study of Urban commuting. In A. Baum, J.E. Singer, S. Valins (Eds.), *Advances in Environmental Psychology* (Vol. 1). Hillsdale: NJ: Erlbaum.

Singh, P. (1987). *Ecology of Urban India* (Vol. 2, pp. 12-13). New Delhi: Ashish Publishing.

Sinha, A.K.P. (1972). Military psychology: A trend report. In S.K. Mitra (Ed.), *A Survey of Research in Psychology*, (pp.238-265). Bombay: Popular Prakashan.

Sinha, D. (1972). Industrial psychology: A trend report. In S.K. Mitra (Ed.), *A Survey of Research in Psychology*, (pp. 175-237). Bombay: Popular Prakashan.

Sinha, M.M. (1972). Physiological and comparative psychology: A trend report. In S.K. Mitra (Ed.), *A Survey of Research in Psychology* (pp. 338-361). Bombay: Popular Prakashan.

Sinha, S.P., Nayyar, P., & Mukherjee, N. (1995). Perception of crowding among children and adolescents. *Journal of Social Psychology*, 135, 263-268.

Skinner, B.F. (1953). *Science and Human Behavior*. New York: Appleton Century Crofts.

Smetana, J., Bridgeman, D.L., & Bridgeman, B. (1978). A field of study of interpersonal distance in early childhood. *Personality and Social Psychology Bulletin*, 4, 309-313.

Smith, M.J., Reinheimer, R.E., & Gabbard-Alley, A. (1981). Crowding, task performance, and communication interaction in youth and old age. *Human Communication Research*, 7, 259-272.

Smith, P.B., & Bond, M.H. (1993). *Social Psychology Across Cultural Analysis and Perspectives*. London: Harvester/Wheatsheaf.

Smith, R.J., & Knowles, E.S. (1979). Affective and cognitive mediators to reactions to spatial invasions. *Journal of Experimental Social Psychology*, 15, 437-452.

Socolow, R.H. (1978). *Saving Energy in the Home*. Cambridge, Mass: Ballinger Publishing Company.

Sommer, R. (1959). Studies in personal space. *Sociometry*, 22, 247-260.

Sommer, R. (1967). Sociofugal space. *American Journal of Sociology*, 72, 654-659.

Sommer, R. (1969). *Personal Space: The Behavioral Basis of Design*. Englewood Cliffs, NJ: Prentice-Hall.

Sommer, R. (1974). *Tight Spaces: Hard Architecture and how to Humanize it*. Englewood Cliff, N. J,: Prentice-Hall.

Sommer, R., & Becker, F.D. (1969). Territorial defense and the good neighbor. *Journal of Personality and Social Psychology*, 11, 85-92.

Southwick, C.H. (1967). An experimental study of intra-group antagonistic behaviour in Rhesus monkeys. *Behavior*, 28, 182-209.

Spangler, D. (1984). *Emergence: The Rebirth of the Sacred*. New York: Dell.

Spencer, C., & Darvizeh, C. (1981). The case for developing a cognitive psychology that does not underestimate the abilities of young children. *Journal of Environmental Psychology*, 1, 21-31.

Spittle, B. (1994). Psychopharmacology of fluoride: A review. *International Clinical Psychopharmacology*, 9, 79-82.

Srivastava, A.K., & Dwivedi, C.B. (1990). Effects of noise and intentional processing retention. *Perspectives in Psychological Researches*, 13, 1-7.

Stamps, A.E. (1994). All buildings great and small: Design review from high-rise to Houses. *Environment and Behavior*, 26, 402-420.

Stea, D. (1969). The measurements of mental maps: An experimental model for studying conceptual spaces. In K.R. Cox & R.G. Golledge (Eds.), *Behavioral Problems in Geography: A Symposium*. Evanston, IL: North-Western University.

Stevens, A., & Coupe, P. (1978). Distortions in judged spatial relations. *Cognitive Psychology*, 10, 422-437.

Stokols, D. (1972). On the distinction between density and crowding: Some implications for future research. *Psychological Review*, 79, 241-254, 275-277.

Stokols, D. (1976). The experience of crowding in primary and secondary environments. *Environment and Behavior*, 1, 8, 49-86.

Stokols, D. (1978). A topology of crowding experience In A. Baum & Y. Epstein (Eds.), *Human Response to Crowding*. Hillsdale, NJ: Erlbaum.

Stokols, D., & Altman, I.(Eds.) (1987). *Handbook of Environmental Psychology*. New York : John Wiley.

Stokols, D., & Novaco, R. (1981). Transportation and well-being: An ecological perspective. In I. Altman, J. Wohlwill, & P. Everett (Eds.), *Human Behavior Environment: Advances in Theory and Research* (Vol 5). New York: Plenum.

Stokols, D., & Ohlig, W. (1975). *The Experience of Crowding under Different Social Climates*. Paper presented at the meeting of the American Psychological Association, Chicago.

Stokols, D., & Shumaker, S.A. (1981). People in places: A transactional view of setting. In J.H. Harvey (Ed.), *Cognition, Social Behavior and the Environment* (pp.441-488), Hillsdale, NJ: Lawrence Erlbaum.

Stokols, D., Rall, M., Pinner, B., & Schopler, J. (1973). Physical, social, and personal determinants of the perception of crowding. *Environment and Behavior,* 5, 87-115.

Sundeen, R.A., & Mathieu, J. (1976). The fear of crime and its consequences among Elderly in three urban communities. *Gerontologist,* 16, 211-219.

Sundstrom, E. (1975). The experimental study of crowding: Effects of room size, intrusion and goal blocking on non-verbal behaviours, self-disclosure, and self-reported stress. *Journal of Personality and Social Psychology,* 32, 645-654.

Sundstrom, E. (1978). A test of equilibrium theory: Effects of topic intimacy and proximity of verbal and non verbal behaviour in pairs of friends and strangers. *Environmental Psychology and Non Verbal Behavior,* 3, 3-16.

Sundstrom, E. (1984). Physical environment and social behaviour. In K. Deaux and L. Wrightsman, *Social Psychology in the Eighties,* 4th edition. (pp. 430-457). Monterey, CA: Brooks/Cole.

Sundstrom, E., & Altman, I. (1976). Interpersonal relationships and personal space: Research review and theoretical model. *Human Ecology,* 4, 47-67.

Sundstrom, E., Bell, P.A., Busby, P.L., & Asmus, C. (1996). Environmental psychology, *Annual Review of Psychology,* 47, 485-512.

Tandon, S.K. (1991). Psychiatric complications of disaster. *Paper presented at the National Seminar on Environment and Behaviour,* Barkatullah University, Bhopal.

Taylor, R.B., & Stough, R. (1978). Territorial cognition: Assessing Altman's typology. *Journal of Personality and Social Psychology,* 36, 418-423.

Taylor, R.B., Gottfredson, S.D., & Brower, S. (1981). Territorial cognitions and social Climate in urban neighbourhoods. *Basic and Applied Social Psychology,* 2, 289-303.

Tedesco, J.F., & Fromme, D.K. (1974). Cooperation, competition and personal space. *Sociometry*, 37, 116-121.

Teilhard, D.C. (1959). *The Phenomenon of Man*. New York: Harper and Row.

Tennis, G.H., & Dabbs, J.M. (1975). Sex, setting and personal space: First grade through college. *Sociometry*, 38, 385-394.

Thompson, D.E., & Aiello, J.R. (1981). *Non-verbal Behavior: Development in Retarded and Non-Retarded Children and Adolescents*. Paper presented at the meeting of the Society for Research in Child Development, Boston.

Thompson, G.O., Raab, G.M., Hepburn, W.S., & Hunter, R. (1989). Blood lead levels of children's behaviour: Results from the Edinburgh Lead study. *Journal of Child Psychology and Psychiatry and Allied Disciplines*, 30, 515-528.

Thompson, W. (1953). *Population Problem* (fourth edition). New York: McGraw-Hill.

Thompson, W.I. (1991). *The American Replacement of Nature*. New York: Doubleday.

Thurston, G. (May, 1997). *Congressional testimony*.

Toffler, A. (1990). *Powershift*. New York: Bantam Books.

Tolman, E.C. (1948). *Purposive Behavior in Animals and Men*. New York: Appleton, Century-Crofts.

Tripathi, H. (1990). *Cognitive and Affective Consequences of Crowding Stress*. Unpublished Doctoral Dissertation, Bhopal university, Bhopal.

Tripathi, L.B. (1997). A theoretical perspective on Man-environment relationship. In U. Jain (Ed.). *Perspectives on Environment and Behaviour*. (pp. 28-37). New Delhi: Har-Anand Publisher.

Tripathi, M. (1993). *Environmental Change and its Psychological Impact: A Case Study of a Village*. Master's thesis in Psychology, Gorakhpur Univesity, Gorakhpur.

Tripathi, P.S.N. (1997). *Pariyavaran Monavighan*, Motilal Banarasidas Publications, Varanasi.

Tripathi, S. (1986). *Effects of Crowding on Human Behavior*. Unpublished Doctoral Dissertation, Kashi Vidhyapeeth, Varanasi.

Trowbridge, C.C. (1948). *Comparative Psychology of Mental Development.* New York: International Universities Press.

Tuan, Y.F. (1974). *Topophilia: A Study of Environmental Perception, Attitude and Values.* Englewoods Cliffs, NJ: Prentice-Hall.

Ugwuegbu, D.C., & Anusiem, A.U. (1982). Effects of stress on interpersonal distance in a simulated interview situation. *Journal of Social Psychology*, 116, 3-7.

Unger, D.G., & Wandersman, A. (1985). The importance of neighbours: The social, cognitive and affective components of neighbouring. *American Journal of Community Psychology*, 13, 139-169.

Vajk, P. (1978). *Doomsday Has Been Cancelled.* Menlo Park, California: Peace Publishers.

Valins, S., & Baum, A. (1973). Residential group size, social interaction and crowding. *Environment and Behavior*, 5, 421-439.

Veitch, R., & Arkkelin, D. (1995). *Environmental psychology: An Interdisciplinary Perspective.* Englewood Cliffs, NJ: Prentice-Hall.

Wallis, A.D. (1981). *Mobile Homes: A Psychological Case Study of Innovation in Housing.* Unpublished Ph.D thesis, University of New York.

Wandersman, A., Florin, P., Chavis, D., Rich, R., & Prestby, J. (1985). Getting together And getting things done. *Psychology Today*, pp. 64-71.

Wapner, S., Kaplan, B., & Cohen, S.B. (1973). An organismic-developmental for understanding transactions of man in environment. *Environment and Behavior*, 5, 255-289.

Watson, O.M. (1970). *Proxemic Behavior: A Cross- Cultural Study.* The Hague, Netherlands: Mouton.

Watson, O.M., & Graves, T.D. (1966). Quantitative research in proxemic behaviour. *American Anthropologist*, 68, 971-985.

Webb, W.M., & Worchel, S. (1993). Prior experience and expectation in the context of Crowding. *Journal of Personality and Social Psychology*, 65, 512-521.

Weir, D. (1987). *The Bhopal Syndrome.* San Francisco, Sierra Club Books.

Weisman, G. (1979). *A Study in Architectural Legibility.* Unpublished doctoral dissertation, University of Michigan.

White, G. (1945). *Human Adjustments to Floods* (Research paper No. 29), Chicago: University of Chicago, Department of Geography.

White, L. (1967). The historical roots of our ecologic crisis. *Science*, 155, March 10.

Whyte, W.H. (1956). *The Organization Man*. Garden City, NY: Doubleday.

Wicker, A. (1969). Size of church membership and members support of church behaviour settings. *Journal of Personality and Social Psychology*, 13, 278-288.

Wicker, A. (1979). *An Introduction to Ecological Psychology*. Monterey, CA: Brooks/Cole.

Wicker, A., & Kauma, C. (1974). Effects of a merger of a small and large organization on members' behaviours and experiences. *Journal of Applied Psychology*, 59, 24-30.

Wicker, A.W. (1979). *An Introduction to Ecological Psychology*. Monterey, CA: Brooks/Cole.

Wicker, A.W. (1987). Behavioral settings reconsidered: Temporal stages, resources, Internal dynamics, context (pp. 613-653). In D. Stokols & I. Altman (Eds.), *Handbook of Environmental Psychology*, New York: John Wiley.

Wicker, A.W., & Kirmeyer, S.L. (1976). From church to laboratory to national park: A programme research on excess and insufficient populations in behaviour settings. In S. Wapner, S.B. Cohen, & B. Kaplan (Eds.), *Experiencing the Environment* (pp. 157-185). New York: Plenum.

Widmayer, S.M., Petersen, L.M., Larner, M., & Carnnhan, S. (1990). Predictors of Haitian-American infant development at twelve months. *Child Development*, 61, 410-415.

Wilson, B.J., & Hopkins, B.L. (1973). The effects of contingent music on the intensity of noise in junior high school economics classes. *Journal of Applied Behavior Analysis*, 6, 269-275.

Wilson, E.O. (1992). *The Diversity of Life*. Cambridge: Harward University Press.

Winsborough, H. (1965). The social consequences of high population density. *Law and Contemporary Problems*, 30, 120-126.

Winston, B.J. (1974). The relationship of awareness to concern for environmental quality among selected high school students. *Dissertation Abstracts International*, 35A, 3412.

Wirth, L. (1938). Urbanism as a way of life. *The American Journal of Sociology*, 44, 1-24.

Wohlwill, J.F. (1973). The environment is not in the head! In W.F.E. Preiser (Ed.), *Environmental Design Research* (Vol. 2) (pp. 166-181). Stroudsburg, PA: Dowden, Hutchinson, & Ross.

Wohlwill, J.F. (1974). Environmental aesthetics: The environment as a source of affect. In I. Altman & J. F. Wohlwill (Eds.), *Human behaviour and environment: Advances in Theory and Research* (Vol 1). New York: Plenum.

Wohlwill, J.F. (1974). Human response to the levels of environmental stimulation. *Human Ecology*, 2, 127-147.

Wolfgang, J., & Wolfgang, A. (1971). Explanation of attitudes via physical interpersonal distance towards the obese, drug users, homosexuals, police, and other marginal figures. *Journal of Clinical Psychology*, 27, 510-512.

Worchel, S., & Teddlie, C. (1976). The experience of crowding: A two- factor theory. *Journal of Personality and Social Psychology*, 34, 30-40.

Worthington, M.E. (1974). Personal space as a function of the stigma effect. *Environment and Behavior*, 6, 289-294.

Yankey, W.C. (1972). Architecture, interaction and social contact: The case of a large scale housing project. In J.G. Wohlwill & D.H. Carson (Eds.), *Environment and the Social Sciences: Perspectives and Application* (pp. 126-153). Washington, DC: American Psychological Association.

Yerkes, R.M., & Dodson, J.D. (1908). The relation of strength of stimulus to rapidity of habit- formation. *Journal of Comparative Neurology and Psychology*, 18, 459-482.

Zeisel, J. (1981). *Inquiry by Design: Tool for Environment-Behavior Research.* Montery, Calif: Brooks/Cole.

Zlutnick, S., & Altman, I. (1972). Crowding and human behaviour. In J.F. Wohlwill & D.H. Carson (Eds.), *Environment and the Social Sciences* (pp. 44-60). Washington, DC: American Psychological Association.

INDEX